Progress in Mathematics
Volume 114

Carlos A. Berenstein
Roger Gay
Alekos Vidras
Alain Yger

Residue Currents and Bezout Identities

Springer Basel AG

Authors:

Carlos A. Berenstein
Mathematics Department &
Institute of Systems Research
University of Maryland
College Park, MD 20742
USA

Alekos Vidras
Research Institute for
Mathematical Sciences
Kyoto University
606 Kyoto
Japan

Roger Gay and
Alain Yger
Centre de Recherche en Mathématiques
Université de Bordeaux I
33405 Talence (Cedex)
France

A CIP catalogue record for this book is available from the Library of Congress, Washington D.C., USA

Deutsche Bibliothek Cataloging-in-Publication Data

Residue currents and bezout identities / Carlos A. Berenstein
... – Basel ; Boston ; Berlin : Birkhäuser, 1993
  (Progress in mathematics ; Vol. 114)
ISBN 978-3-0348-9680-1   ISBN 978-3-0348-8560-7 (eBook)

DOI 10.1007/978-3-0348-8560-7

NE: Berenstein, Carlos A.; GT
© 1993 Springer Basel AG
Originally published by Birkhäuser Verlag Basel in 1993
Softcover reprint of the hardcover 1st edition 1993

Camera-ready copy prepared by the authors
Printed on acid-free paper produced of chlorine-free pulp
ISBN 978-3-0348-9680-1

9 8 7 6 5 4 3 2 1

To our families, for their infinite patience.

In memoriam Miguel Herrera

Table of Contents

Introduction

A very primitive form of this monograph has existed for about two and a half years in the form of handwritten notes of a course that Alain Yger gave at the University of Maryland. The objective, all along, has been to present a coherent picture of the almost mysterious role that analytic methods and, in particular, multidimensional residues, have recently played in obtaining effective estimates for problems in commutative algebra [71;5]*

Our original interest in the subject rested on the fact that the study of many questions in harmonic analysis, like finding all distribution solutions (or finding out whether there are any) to a system of linear partial differential equations with constant coefficients (or, more generally, convolution equations) in $\mathbb{R}^n$, can be translated into interpolation problems in spaces of entire functions with growth conditions. This idea, which one can trace back to Euler, is the basis of Ehrenpreis's Fundamental Principle for partial differential equations [37;5], [56;5], and has been explicitly stated, for convolution equations, in the work of Berenstein and Taylor [9;5] (we refer to the survey [8;5] for complete references.) One important point in [9;5] was the use of the Jacobi interpolation formula, but otherwise, the representation of solutions obtained in that paper were not explicit because of the use of $\bar{\partial}$-methods to prove interpolation results. (The same was true in [37;5], [56;5], which should be compared to Chapter 5, §3, below.) The problem of deconvolution, and its possible applications to image and signal processing, brought to the forefront the need to find explicit interpolation formulas in $\mathbb{C}^n$ to solve the so called analytic Bezout identities (an engineering name), which are needed to find deconvolution kernels. One solution was given in [10;5], where such formulas were given as series, the general term of which was given by the Jacobi interpolation formula. It took some prodding by friends from the Institute of Systems Research , to realize the obvious, that the series from [10;5] truncate when one is dealing with polynomials. This observation is one of the major components of our recent work on effective bounds for solutions to the algebraic Bezout equation and the membership problem for polynomial ideals, and it will be systematically exploited in this book.

Before we proceed further, let us recall that the Bezout equation is simply an equation of the form

$$f_1 g_1 + \cdots + f_m g_m = 1 \,,$$

where $f_1, \ldots, f_m$ are, for example, polynomials (or entire functions with growth conditions), and one looks for polynomial solutions $g_1, \ldots, g_m$ (or entire functions with the growth conditions similar to those satisfied by the f_j.) In the

---

* The symbol [71;5] refers to item 71 in the References to Chapter 5.

algebraic case, one would like to find also solutions with "good" bounds on the degrees and "size" of the coefficients. This a priori knowledge plays a role in complexity estimates in Computer Algebra and it has applications in Transcendental Number Theory [71;5]. The first person to obtain good estimates for the degrees in the algebraic Bezout equation was Dale Brownawell, and our work on this subject owes a lot to his encouragement.

The other ingredient in our work on Bezout identities is the observation that the Jacobi interpolation formulas have something to do with residues. In fact, Jacobi introduce them (and, apparently, also the concept of the Jacobian determinant) [8;4], for the purpose of generalizing to several variables the theorem of Abel that the sum of residues, over the complex plane, of a rational function vanishes if the degree of the denominator exceeds that of the numerator by 2. (We generalize further Jacobi's result in Theorem 4.13.) The residue that appears in Jacobi's work is the following: let us say, that P, Q, and R are functions of two complex variables x, y, holomorphic in a neighborhood of $x = y = 0$, and that $(0,0)$ is the only common zero of P and Q, then Jacobi's residue is

$$\lim_{\epsilon \to 0+} \int_{\substack{|P|=\epsilon \\ |Q|=\epsilon}} \frac{R(x,y)}{P(x,y)Q(x,y)} dx \wedge dy \,.$$

Let us just mention here manifold applications of this concept can be found in [16;3], [1;2], [2;4] (and even a forthcoming book of Aizenberg with applications to quantum chemistry!) and refer the reader to Chapters 2 and 3 for more details and complete references. Because in our work we need to use integral representation formulas of the Henkin type for holomorphic functions, and, contrary to the one variable case, they require non-holomorphic kernels, and these are the integrals we compute "by" residues, we need to consider residue currents, i.e., R is allowed to be only smooth (not necessarily holomorphic.) In that case, even the existence of the above limit is questionable. The existence of residue currents and many of their properties are based on the work of Dolbeault and of the late Argentinian mathematician Miguel Herrera (see [13;3], [38;3].) Inspired in part by Passare's thesis and by the work of Bernstein on $\mathcal{D}$-modules, we took a different approach to residue currents, which works when one is in the situation of Jacobi, i.e., P, Q define a complete intersection, namely, the residue current is related to the analytic continuation of the distribution-valued holomorphic function

$$\mu \mapsto |P(x,y)Q(x,y)|^{2\mu} \,,$$

originally defined only for $Re\,\mu > 0$. It is this observation that, as we show in this monograph, ties together all the different components of our work on Bezout identities. At least, we hope to convince the reader that such is the case, and therefore, it merits his attention as well as interest in solving many of the questions we have left open.

To conclude, let us mention that the research that led to this monograph was supported over the years by grants from several organizations: the National Science Foundation, the Air Force Office of Scientific Research, the Army Research Office, the National Security Agency, and the Conseil Nationale de la Recherche, to all of them we give our heartfelt thanks. We would also like to express our gratitude to M. Elkadi, who corrected many of our mistakes, as well as to many other friends, who encouraged us to complete this monograph and gave us much needed advice, especially, A. Dickenstein and P. Pedersen.

Bethesda, MD
June 7, 1993

Chapter 1
Residue Currents in one Dimension
Different Approaches

§1 Residue attached to a holomorphic function

Let f be an holomorphic function in a domain $U \subset \mathbb{C}$. We assume that $f \not\equiv 0$, so that the zero set of f is a discrete subset of U. Let K be a relatively compact subset of U and $\phi \in \mathcal{D}(U)$ such that $\text{supp}(\phi) \subset K$, our first goal is to study the expression:

$$I(\phi, \epsilon) = \frac{1}{2\pi i} \int_{|f|=\epsilon} \frac{\phi(\zeta)}{f(\zeta)} d\zeta \tag{1.1}$$

and more precisely what happens if ϵ tends to 0.

We need first to point out that the definition of $I(\phi, \epsilon)$ for arbitrary ϵ is not so clear: the integration over $\{|f| = \epsilon\}$ can be performed when $\{|f| = \epsilon\}$ is a manifold or at least a piecewise smooth curve in a neighborhood of $\text{supp}(\phi)$; moreover for the definition (1.1) to be complete, we have to give an orientation to the set $\{|f| = \epsilon\}$. As simplexes and chains will be introduced in Chapter 2, we overcome these technical difficulties here by introducing a partition of unity as follows:

Let $\alpha \in K$; in a disk around α, one can write $f(\zeta) = (\zeta - \alpha)^{m(\alpha)} u_\alpha(\zeta)$ where u_α is a nonvanishing holomorphic function in this disk Δ_α; this allows us to express f in Δ_α as $f(\zeta) = (\zeta - \alpha)^{m(\alpha)} \exp(\phi_\alpha(\zeta))$ where ϕ_α is holomorphic in Δ_α. When $m(\alpha) > 0$, the transformation $\zeta \mapsto w$, defined by

$$\zeta - \alpha = w \exp\left(-\frac{\phi_\alpha(\zeta)}{m(\alpha)}\right),$$

is a local diffeomorphism θ_α between the disk Δ_α in the $\mathbb{C}_\zeta$ plane and some neighborhood of 0 in the $\mathbb{C}_w$ plane. We cover K with a finite number of such disks $\Delta_{\alpha_1}, \Delta_{\alpha_2}, \ldots, \Delta_{\alpha_M}$ and introduce a partition of unity $(\phi_{\alpha_i})_{i=1,\ldots,M}$ associated to such a covering (we recall that $\text{supp}(\phi_{\alpha_i}) \subseteq \Delta_{\alpha_i}$ and that $\sum_{j=1}^{j=M} \phi_{\alpha_j} \equiv 1$ in a neighborhood of K). We rewrite (1.1) as :

$$I(\phi, \epsilon) = \frac{1}{2\pi i} \sum_{j=1}^{M} \int_{|f|=\epsilon} \frac{\phi_{\alpha_j}(\zeta)\phi(\zeta)}{f(\zeta)} d\zeta$$

For ϵ small enough, the set $\{|f| = \epsilon\} \cap \text{supp}(\phi_{\alpha_j})$ is empty as soon as $m(\alpha_j) = 0$, therefore, for ϵ small enough,

$$I(\phi, \epsilon) = \frac{1}{2\pi i} \sum_{\substack{1 \le j \le M \\ f(\alpha_j)=0}} \int_{|f|=\epsilon} \frac{\phi_{\alpha_j}(\zeta)\phi(\zeta)}{f(\zeta)} d\zeta.$$

Let us remark that the integrals $\int_{|f|=\epsilon} (\phi_{\alpha_j}(\zeta)\phi(\zeta)/f(\zeta))\,d\zeta$ can be defined without ambiguity using the diffeomorphisms θ_{α_j}, namely :

$$\int_{|f|=\epsilon} \frac{\phi_{\alpha_j}(\zeta)\phi(\zeta)}{f(\zeta)}\,d\zeta = \int_{|w|=\epsilon^{1/m(\alpha_j)}} \frac{(\phi_{\alpha_j}\phi)(\theta_{\alpha_j}^{-1})(w)}{w^{m(\alpha_j)}}(\theta_{\alpha_j}^{-1})'(w)\,dw.$$

The orientation of $\{|f| = \epsilon\}$ is induced by the positive orientation of the circle $\{|w| = \epsilon^{1/m(\alpha_j)}\}$ in the $\mathbb{C}_w$ plane. We may now state our first result about the behaviour of $I(\phi, \epsilon)$ when $\epsilon \to 0$.

Proposition 1.1. *Let $f \not\equiv 0$ be an holomorphic function in a domain $U \subset \mathbb{C}$. For every $\phi \in \mathcal{D}(U)$, the limit*

$$T_f(\phi d\zeta) := \lim_{\epsilon \to 0}(I(\phi, \epsilon)), \tag{1.2}$$

exists and defines a $(0,1)-$current in U. Furthermore, for any $K \subset\subset U$, there exists a positive constant $C(K)$ such that, if $\mathrm{supp}(\phi) \subset K$, then

$$|T_f(\phi d\zeta)| \le C(K) \max_{\substack{f(\alpha)=0 \\ \alpha \in K \\ p < m(\alpha)}} |\frac{\partial^p}{\partial \zeta^p}\phi(\alpha)|.$$

Definition 1.2. *The current T_f defined by (1.2) is called the residue current attached to f.*

Proof. From the remarks above, it is enough to study

$$\int_{|f|=\epsilon} \frac{\phi_\alpha(\zeta)\phi(\zeta)}{f(\zeta)}\,d\zeta\,,$$

when α is a zero of f. Let $m(\alpha) = m$, for the sake of simplicity. We have

$$\int_{|f|=\epsilon} \frac{\phi_\alpha(\zeta)\phi(\zeta)}{f(\zeta)}\,d\zeta = \int_{|w|=\epsilon^{1/m}} \frac{(\phi_\alpha\phi)(\theta_\alpha^{-1}(w))}{w^m}(\theta_\alpha^{-1})'(w)\,dw. \tag{1.3}$$

We expand $(\phi_\alpha\phi)(\theta_\alpha^{-1})(w)(\theta_\alpha^{-1})'(w)$ as a Taylor series in the variables w, $\bar{w}$ near the origin . Thus,

$$\psi_\alpha = (\phi_\alpha\phi)(\theta_\alpha^{-1}(w))(\theta_\alpha^{-1})'(w) = \sum_{k+l \le m-1} \frac{1}{k!l!}(\frac{\partial^{k+l}}{\partial^k\bar{\partial}^l\zeta}\psi_\alpha)_{w=0}\, w^k\bar{w}^l + o(|w|^{m-1}).$$

For $k + l < m - 1$, we have

$$\int_{|w|=\epsilon^{1/m}} \frac{w^k\bar{w}^l}{w^m}\,dw = i\int_0^{2\pi} e^{i(k-l-m+1)\theta}\,d\theta = 0.$$

We can write (1.3) as

$$\frac{2\pi i}{(m-1)!}\Big(\frac{\partial^{m-1}}{\partial w^{m-1}}\psi_\alpha\Big)_{w=0} + \int_{|w|=\epsilon^{1/m}} \frac{o(|w|^{m-1})}{w^m}\,dw.$$

This implies that the limit of (1.3) when $\epsilon \to 0$ is equal to

$$\frac{2\pi i}{(m-1)!}\Big(\frac{\partial^{m-1}}{\partial w^{m-1}}\psi_\alpha\Big)_{w=0}.$$

The proposition follows from this observation. $\square$

Remark 1.3. If we assume ϕ to be holomorphic near the zero set of f, we can compute $I(\phi,\epsilon)$ easily, in fact, for ϵ sufficiently small, $\{|f| = \epsilon\}$ is a collection of Jordan curves surrounding the points α in $\{f = 0\} \cap K$ (these curves are images of $|w| = \epsilon^{1/m(\alpha)}$ by the diffeomorphisms θ_α). From the usual residue theorem, we have

$$I(\phi,\epsilon) = \sum_{\alpha \in \{f=0\}\cap K} \frac{1}{(m(\alpha)-1)!}\Big(\frac{\partial^{m(\alpha)-1}}{\partial \zeta^{m(\alpha)-1}}[(\zeta-\alpha)^{m(\alpha)}\frac{\phi(\zeta)}{f(\zeta)}]\Big)_{\zeta=\alpha}.$$

It is now clear from the proof of our proposition that if $\phi \in \mathcal{D}(U)$, then

$$T_f(\phi d\zeta) = T_f\Big(\sum_{\alpha \in K} \tilde{\phi}_\alpha \psi_\alpha d\zeta\Big),$$

where ψ_α is an element in $\mathcal{D}(\Delta_\alpha)$ such that $\psi_\alpha \equiv 1$ near K and $\tilde{\phi}_\alpha$ is the holomorphic part of ϕ, that is,

$$\tilde{\phi}_\alpha(\zeta) = \sum_{k=0}^{m(\alpha)} \frac{1}{k!}\Big(\frac{\partial^k \phi}{\partial \zeta^k}\Big)_{\zeta=\alpha}(\zeta-\alpha)^k.$$

Therefore, whenever $\phi \in \mathcal{D}(U)$ and $\mathrm{supp}(\phi) \subset K$,

$$T_f(\phi d\zeta) = \sum_{\alpha \in \{f=0\}\cap K} \frac{1}{(m(\alpha)-1)!}\Big(\frac{\partial^{m(\alpha)-1}}{\partial \zeta^{m(\alpha)-1}}[(\zeta-\alpha)^{m(\alpha)}\frac{\phi(\zeta)}{f(\zeta)}]\Big)_{\zeta=\alpha}. \qquad (1.4)$$

This remark will be crucial in the multidimensional case later on.

Remark 1.4. Suppose K' is a compact set bounded by Jordan curves such that $K \subset \mathrm{int}(K') \subset K' \subset U$ and $f \neq 0$ on $\partial K'$. Using Stokes's theorem, whenever $\phi \in \mathcal{D}(U)$, $\mathrm{supp}\,\phi \subset K$ and ϵ is sufficiently small, we have

$$\int_{\partial K'} \frac{\phi}{f}\,d\zeta - \int_{|f|=\epsilon} \frac{\phi}{f}\,d\zeta = -\int_{|f|=\epsilon} \frac{\phi}{f}\,d\zeta = \int_{|f|\geq\epsilon} \frac{\bar{\partial}(\phi\,d\zeta)}{f}.$$

Since

$$T_f(\phi d\zeta) = \lim_{\epsilon \to 0} \frac{1}{2\pi i} \int_{|f|=\epsilon} \frac{\phi}{f} d\zeta$$

$$= \lim_{\epsilon \to 0} \frac{-1}{2\pi i} \int_{|f| \geq \epsilon} \frac{\bar{\partial}(\phi \, d\zeta)}{f}$$

$$= \lim_{\epsilon \to 0} \frac{1}{2\pi i} < \bar{\partial}(\frac{1}{f}\chi_{|f| \geq \epsilon}), \phi d\zeta >,$$

it is natural, in the language of currents, to use the notation

$$< \bar{\partial}\frac{1}{f}, \phi d\zeta >:= T_f(\phi d\zeta).$$

§2 Some other approaches to the residue current

As before, we consider a function f holomorphic in a domain $U \subset \mathbb{C}$, $f \not\equiv 0$; K is a relatively compact subset of U and ϕ is an element of $\mathcal{D}(U)$ whose support lies in K. Following Gelfand [5], we introduce the function of the complex parameter λ

$$\lambda \mapsto J(\lambda, \phi) = \frac{1}{2\pi i}\lambda \int_{\mathbb{C}} |f(\zeta)|^{2(\lambda-1)}\overline{\partial f(\zeta)} \wedge \phi(\zeta)d\zeta. \qquad (1.5)$$

It follows from Lebesgue's dominated convergence theorem, together with Morera's theorem, that the function $\lambda \mapsto J(\lambda, \phi)$ is holomorphic in the half-plane Re $\lambda > 1$. In fact, we can analytically continue this function to some right half-plane Re $\lambda > -\eta$, $\eta > 0$ as a consequence of the following proposition.

Proposition 1.5. *Let f, U, K, and ϕ be as earlier. Then, the function $\lambda \mapsto J(\lambda, \phi)$ defined by (1.5) when Re $\lambda > 1$ can be analytically continued as a holomorphic function in the half-plane*

$$\text{Re } \lambda > -[\max\{m(\alpha) : \alpha \in K, \, f(\alpha) = 0\}]^{-1}.$$

Proof. We need the partition of unity introduced in §1 (keeping the same notations), we have

$$J(\lambda, \phi) = \sum_{j=1}^{M} \frac{1}{2\pi i}\lambda \int |f(\zeta)|^{2(\lambda-1)}\overline{\partial f(\zeta)} \wedge \phi\phi_{\alpha_j}(\zeta)d\zeta$$

$$= J_1(\lambda, \phi) + J_2(\lambda, \phi),$$

where

$$J_1(\lambda, \phi) = \sum_{\substack{1 \leq j \leq M \\ f(\alpha_j)=0}} \frac{1}{2\pi i}\lambda \int |f(\zeta)|^{2(\lambda-1)}\overline{\partial f(\zeta)} \wedge \phi\phi_{\alpha_j}(\zeta)d\zeta,$$

and, as we shall always do, we have suppressed the domain of integration since it is the whole plane. When $f(\alpha_j) \neq 0$, f does not vanish on the support of ϕ_{α_j}, then $\lambda \mapsto J_2(\lambda, \phi)$ can be extended as an entire function of λ. We are led to study a function of the form

$$J_{1,\alpha_j}(\lambda, \phi) = \frac{1}{2\pi i}\lambda \int |f(\zeta)|^{2(\lambda-1)}\overline{\partial f(\zeta)} \wedge \phi\phi_{\alpha_j}(\zeta)d\zeta$$

where α_j is a zero of f. When Re $\lambda > 2$, one can write $|f|^{2\lambda}/f = |f|^{2(\lambda-1)}\bar{f}$. This fonction is in $\mathcal{C}^1(U)$ and we have

$$\bar{\partial}[\frac{|f|^{2\lambda}}{f}\phi\phi_{\alpha_j}\,d\zeta] = \lambda|f(\zeta)|^{2(\lambda-1)}\overline{\partial f(\zeta)} \wedge \phi\phi_{\alpha_j}(\zeta)d\zeta + \frac{|f|^{2\lambda}}{f}\bar{\partial}(\phi\phi_{\alpha_j}) \wedge d\zeta.$$

Using Stokes's theorem (which will be one of our major tools in most of the proofs), we deduce that

$$\int d[\frac{|f|^{2\lambda}}{f}\phi\phi_{\alpha_j}\,d\zeta] = \int \bar{\partial}[\frac{|f|^{2\lambda}}{f}\phi\phi_{\alpha_j}\,d\zeta] = 0.$$

This is a consequence of the fact that $\phi\phi_{\alpha_j} \in \mathcal{D}(U)$; therefore, for Re $\lambda > 2$ we have

$$J_{1,\alpha_j}(\lambda, \phi) = -\frac{1}{2\pi i}\int \frac{|f|^{2\lambda}}{f}\bar{\partial}(\phi\phi_{\alpha_j}) \wedge d\zeta. \qquad (1.6)$$

The Proposition 1.5 will then be a consequence of the following lemma.

Lemma 1.6. *Let f, U, K, and ϕ be as in Proposition 1.5. Let $N \in \mathbb{N}^*$, then the function $\lambda \mapsto J_{(N)}(\lambda, \phi)$, defined for Re $\lambda > N/2$ by*

$$J_{(N)}(\lambda, \phi) = \int \frac{|f|^{2\lambda}}{f^N}\phi d\bar{\zeta} \wedge d\zeta\,,$$

can be continued as a holomorphic function to the half-plane

$$\text{Re } \lambda > -[\max\{m(\alpha) : \alpha \in K, f(\alpha) = 0\}]^{-1}.$$

Proof of Lemma 1.6. We remark first that it is enough to prove the lemma for $N = 1$. In fact, for Re λ sufficiently large, we have

$$J_{(N)}(\lambda, \phi) = \tilde{J}_{(1)}(\frac{\lambda}{N}, \phi),$$

where f is replaced by f^N in the definition of $\tilde{J}_{(1)}$ (compared to that of $J_{(1)}$). In the proof we shall use the following formal identity

$$(\frac{d}{dx})^m[(x-a)^{m(\lambda+1)}] = (m\lambda+m)(m\lambda+m-1)\ldots(m\lambda+1)(x-a)^{m\lambda}. \qquad (1.7)$$

Introducing a partition of unity allows us to assume that on $\text{supp}(\phi)$, we have $f(\zeta) = (\zeta - \alpha)^m u_\alpha(\zeta)$, where u_α is a nonvanishing holomorphic function. Therefore,

$$J_{(N)}(\lambda, \phi) = \int \frac{|\zeta - \alpha|^{2m\lambda}}{(\zeta - \alpha)^m} |u_\alpha(\zeta)|^{2\lambda} \frac{\phi(\zeta)}{u_\alpha(\zeta)} d\bar{\zeta} \wedge d\zeta$$

$$= \int \frac{|\zeta - \alpha|^{2m\lambda}}{(\zeta - \alpha)^m} \theta_{\alpha,\lambda}(\zeta) d\bar{\zeta} \wedge d\zeta \qquad (1.8)$$

where $(\lambda, \zeta) \mapsto \theta_{\alpha,\lambda}(\zeta)$ is an entire function of λ, which is C^∞ as a function of ζ. When $\text{Re } \lambda \gg 0$, we transform (1.8) by repeated integration by parts with respect to the variable $\bar{\zeta}$ (in other words, by repeated use of Stokes's theorem.) More precisely, we use the following consequence of the identity (1.7)

$$\bar{\partial}\left[\frac{|\zeta - \alpha|^{2m\lambda}}{(\bar{\zeta} - \bar{\alpha})^p}\right] = (m\lambda - p)|\zeta - \alpha|^{2m\lambda} \frac{1}{(\bar{\zeta} - \bar{\alpha})^{p+1}}.$$

This formal identity follows from (1.7) since $|\zeta - \alpha|^2 = (\zeta - \alpha)(\bar{\zeta} - \bar{\alpha})$. It implies that for $p \in \mathbb{Z}$ and $\text{Re } \lambda > \gamma(p) \gg 0$

$$(m\lambda - p)\int |\zeta - \alpha|^{2m\lambda} \frac{1}{(\bar{\zeta} - \bar{\alpha})^{p+1}} \frac{1}{(\zeta - \alpha)^m} \theta_{\alpha,\lambda}(\zeta) d\bar{\zeta} \wedge d\zeta$$

$$= -\int |\zeta - \alpha|^{2m\lambda} \frac{1}{(\bar{\zeta} - \bar{\alpha})^p} \frac{1}{(\zeta - \alpha)^m} \frac{\partial}{\partial \bar{\zeta}}[\theta_{\alpha,\lambda}(\zeta)] d\bar{\zeta} \wedge d\zeta. \quad (1.9)$$

Applying successively (1.9) with $p = -1$, $p = -2$, etc., we obtain

$$J_{(N)}(\lambda, \phi) = \frac{(-1)^m}{\prod\limits_{k=1}^{m}(m\lambda + k)} \int |\zeta - \alpha|^{2m\lambda}\left(\frac{\bar{\zeta} - \bar{\alpha}}{\zeta - \alpha}\right)^m \frac{\partial^m}{\partial \bar{\zeta}^m}[\theta_{\alpha,\lambda}(\zeta)] d\bar{\zeta} \wedge d\zeta. \quad (1.10)$$

Clearly, the right hand side of (1.10) defines a function of λ that is holomorphic in $\text{Re } \lambda > 0$. If we use polar coordinates, i.e., $\zeta - \alpha = \rho e^{i\theta}$, it is not hard to see that this function is holomorphic in $\text{Re}(2m\lambda + 1) > -1$, i.e., $\text{Re } \lambda > -1/m$ as claimed. Thus, Lemma 1.6 and Proposition 1.5 are proved. $\qquad \square$

It turns out that Proposition 1.5 provides another way of defining the current $\bar{\partial}(1/f)$, as it is shown by the following proposition.

Proposition 1.7.   *Let f, U, K, and ϕ be as in the previous proposition. Then,*

$$J(0, \phi) = <\bar{\partial}\frac{1}{f}, \phi d\zeta> .$$

Proof. As above, we use a partition of unity subordinated to the family $\{\Delta_\alpha\}$ and to K. We claim that for any α, among $\alpha_1, \ldots, \alpha_M$, such that $f(\alpha) = 0$, we have

$$\lambda \int |f(\zeta)|^{2(\lambda-1)}\overline{\partial f(\zeta)} \wedge (\phi\phi_\alpha)(\zeta) d\zeta = \lambda \int_0^\infty s^{\lambda-1}\left[\int_{|f|^2 = s} \frac{\phi\phi_\alpha}{f} d\zeta\right]ds. \quad (1.11)$$

To verify the claim, for $m := m(\alpha) \neq 0$, we use the diffeomorphism θ_α to reduce the problem to the $\mathbb{C}_w$ plane, with $w \mapsto w^m$ instead of the function f. In this case, (1.11) becomes

$$m\lambda \int_{\mathbb{C}} |w|^{2m(\lambda-1)} \bar{w}^{m-1} d\bar{w} \wedge \psi_\alpha(w) dw = \lambda \int_0^\infty s^{\lambda-1} [\int_{|w|^2=s} \frac{\psi_\alpha(w)}{w^m} dw] ds.$$
(1.12)

Checking (1.12) can be done using polar coordinates.

For any such α, we know from Section 1 that the function

$$\rho_\alpha : s \mapsto \frac{1}{2\pi i} \int_{|f|^2=s} \frac{\phi_\alpha(\zeta)\phi(\zeta)}{f(\zeta)} d\zeta$$

is continuous on $[0,\infty)$ and its value at $s = 0$ is $< \bar{\partial}(1/f), \phi\phi_\alpha d\zeta >$. Let $A > 0$ be such that $\psi_\alpha \equiv 0$ for $|w|^2 > A$. Then, $\rho_\alpha(s) = 0$ for $s^2 > A$, hence, for $\text{Re } \lambda > 0$,

$$\lambda \int_0^\infty s^{\lambda-1} \rho_\alpha(s) ds - \rho_\alpha(0) = \lambda \int_0^A s^{\lambda-1}(\rho_\alpha(s) - \rho_\alpha(0)) ds + (\lambda \int_0^A s^{\lambda-1} ds - 1)\rho_\alpha(0)$$

$$= \lambda \int_0^A s^{\lambda-1}(\rho_\alpha(s) - \rho_\alpha(0)) ds + (A^\lambda - 1)\rho_\alpha(0). \quad (1.13)$$

For any $0 < \eta < A$ and any such λ we also have the following inequality

$$|\lambda \int_0^A s^{\lambda-1}(\rho_\alpha(s) - \rho_\alpha(0)) ds| \leq \eta^\lambda \max_{|s| \leq \eta} |\rho_\alpha(s) - \rho_\alpha(0)| + 2(\max |\rho_\alpha|)(A^\lambda - \eta^\lambda).$$
(1.14)

It follows from (1.13) and (1.14) that

$$\lim_{\lambda \to 0} (\lambda \int_0^\infty s^{\lambda-1} \rho_\alpha(s) ds) = \rho_\alpha(0) = < \bar{\partial}\frac{1}{f}, \phi\phi_\alpha d\zeta > .$$

This concludes the proof of the above proposition, since all the contributions of to the value of $J(\lambda, \phi)$ at $\lambda = 0$ come from the values at the same point of the functions $\lambda \mapsto J(\lambda, \phi\phi_{\alpha_j})$, for $m(\alpha_j) \neq 0$. $\qquad\square$

In the proof of Proposition 1.7 we have used a Tauberian argument to recover in a different way the action of the residue current. In fact, the function $\lambda \mapsto J(\lambda, \phi\phi_\alpha)$ is the Mellin transform of the function ρ_α. There are other Tauberian methods one can use to obtain the residue current. Here is an example.

Proposition 1.8. *Let f, U, K, and ϕ be as in the previous propositions. Then,*

$$\lim_{\tau \to 0+} \frac{1}{2\pi i} \int \frac{\tau \overline{\partial} f \wedge \phi d\zeta}{(\tau + |f|^2)^2} = < \overline{\partial}\frac{1}{f}, \phi d\zeta > .$$

Proof. Repeating the argument of the last proof and choosing A as above, we show that for α such that $f(\alpha) = 0$,

$$\int \frac{\tau \overline{\partial} f \wedge \phi \phi_\alpha d\zeta}{(\tau + |f|^2)^2} = \int_0^\infty \frac{\tau}{(\tau + s)^2} \left(\int_{|f|^2 = s} \frac{\phi_\alpha(\zeta)\phi(\zeta)}{f(\xi)} d\zeta \right) ds$$

$$= \int_0^A \frac{\tau}{(\tau + s)^2} \left(\int_{|f|^2 = s} \frac{\phi_\alpha(\zeta)\phi(\zeta)}{f(\zeta)} d\zeta \right) ds$$

$$= \int_0^A \frac{\tau}{(\tau + s)^2} \rho_\alpha(s) ds.$$

As before, if $0 < \eta < A$,

$$\left| \int_0^A \frac{\tau}{(\tau + s)^2} \rho_\alpha(s) ds - \rho_\alpha(0) \right| \le \int_0^A \frac{\tau |\rho_\alpha(s) - \rho_\alpha(0)|}{(\tau + s)^2} ds + |\rho_\alpha(0)| (1 - \frac{\tau}{\tau + A})$$

$$\le \int_0^\eta \frac{\tau |\rho_\alpha(s) - \rho_\alpha(0)|}{(\tau + s)^2} ds + 2(\max |\rho_\alpha|) \frac{\tau(A - \eta)}{(\tau + \eta)(\tau + A)} + |\rho_\alpha(0)| (1 - \frac{\tau}{\tau + A}).$$

Choosing η small enough we conclude that

$$\lim_{\tau \to 0} \frac{1}{2\pi i} \int \frac{\tau \overline{\partial} f \wedge \phi \phi_\alpha d\zeta}{(\tau + |f|^2)^2} = < \overline{\partial}\frac{1}{f}, \phi \phi_\alpha d\zeta > .$$

Adding these identities, we obtain the desired result.   □

§3 Some variants of the classical Pompeiu formula

We first recall in this section the classical formula due to Cauchy and Pompeiu.

Proposition 1.9 *Let U be any bounded domain in $\mathbb{C}$ whose boundary is a piecewise C^1 Jordan curve and let $f \in C^1(U)$, then the following representation formula holds for $z \in U$*

$$f(z) = \frac{1}{2\pi i} \left[\int_{\partial U} \frac{f(\zeta)}{\zeta - z} d\zeta - \int_U \frac{\partial f}{\partial \overline{\zeta}}(\zeta) \frac{d\overline{\zeta} \wedge d\zeta}{\zeta - z} \right]. \tag{1.15}$$

Proof. This statement is just a rephrasing of the known result from the theory of distributions in R^2 (see e.g., [5])

$$\frac{\partial}{\partial\bar{\zeta}}\left(\frac{1}{\pi\zeta}\right) = \delta_{(0,0)}. \tag{1.16}$$

In other words, $1/\pi\zeta$ is a fundamental solution to the differential operator $\partial/\partial\bar{\zeta}$. Namely, let $(\rho_\epsilon)_{\epsilon>0}$ be a regularization of $\delta_{(0,0)}$, then

$$\frac{1}{2\pi i}\int_{\partial U}\frac{f(\zeta)}{\zeta - z}d\zeta = \lim_{\epsilon\to 0}\frac{1}{2\pi i}\int_{\partial U}f(\zeta)[\frac{1}{t-z}*\rho_\epsilon(t)](\zeta)d\zeta$$

$$= \lim_{\epsilon\to 0}\frac{1}{2\pi i}\int_U \bar{\partial}[f(\zeta)(\frac{1}{t-z}*\rho_\epsilon(t))(\zeta)]\wedge d\zeta$$

$$= \frac{1}{2\pi i}\int_U \frac{\partial f}{\partial\bar{\zeta}}(\zeta)\frac{1}{\zeta - z}d\bar{\zeta}\wedge d\zeta + <\frac{\partial}{\partial\bar{\zeta}}[\frac{1}{\pi(\zeta - z)}], f>.$$

Therefore, (1.15) follows from (1.16). Note that a direct proof of this result can be found, e.g., in [3]. □

We want to construct "weighted" versions of this formula. For that purpose, let us introduce some open set ω, $\omega \subset U$, M functions of two variables $q_1, \ldots, q_M$ belonging to $C^1(\bar{\omega}\times\bar{U})$, and M functions $G_1, \ldots, G_M$, each G_k defined and holomorphic in a neighborhood of the image of $\bar{\omega}\times\bar{U}$ by the mapping

$$(z, \zeta) \mapsto \Phi_k(z, \zeta) := 1 + q_k(z, \zeta)(z - \zeta)$$

and satisfying $G_k(1) = 1$. For the sake of simplicity, we denote

$$\Gamma_k^{(\alpha)}(z, \zeta) = \frac{d^\alpha}{dt^\alpha}G_k|_{t=\Phi_k(z,\zeta)}, \; z\in\bar{\omega}, \; \zeta\in\bar{U}.$$

The weighted version of Pompeiu's formula (1.15) is given by the following proposition.

Proposition 1.10. *Let U, f be as in the statement of the previous proposition. Let $\omega \subset U$ and $q_1, \ldots, q_M, G_1, \ldots, G_M$ be as in the preceding considerations. For $z \in \omega$, we have*

$$f(z) = \frac{1}{2\pi i}[\int_{\partial U}\frac{f(\zeta)\prod_{k=1}^{M}\Gamma_k^{(0)}(z,\zeta)}{\zeta - z}d\zeta + \sum_{j=1}^{M}\int_U f\Gamma_j^{(1)}\prod_{\substack{k=1\\k\neq j}}^{M}\Gamma_k^{(0)}(z,\zeta)\bar{\partial}_\zeta q_j\wedge d\zeta].$$

$$\tag{1.17}$$

Proof. Note that in (1.17) we have left implicit those variables that are evident from the context. It is a practice that we will carry on throughout this book.

The proof of (1.17) is immediate. The key point consists in applying (1.15) to the function

$$g_z : \zeta \mapsto f(\zeta) \prod_{k=1}^{M} \Gamma_k^{(0)}(z, \zeta),$$

where z is kept fixed in ω. It is clear from the hypothesis that $G_k(1) = 1$, that $g_z(z) = f(z)$. $\hfill\square$

The point which is worthwile noticing here is that the singularity in the second term of the right handside of (1.15) disappears in (1.17); this is a nice phenomenon, which occurs in the case of one variable, due to the fact that the division by $z - \zeta$ is elementary in this case. We shall try to find an analogue of the trivial identity $1 = (\frac{1}{z-\zeta})(z - \zeta)$ in the multidimensional case later on.

§4 Some applications of Pompeiu's formulas. Local results

In this section, we consider m functions $f_1, \ldots, f_m$, which are holomorphic in a neighborhood of the closed unit disk $\bar{D}$ and have no common zeros on $|\zeta| = 1$. Certainly the unit disk can be replaced by some more general domain, for example a bounded connected component of the set where the functions $f_1, \ldots, f_m$ are simultaneously small (see [1] for problems of this kind.) Our goal here is to solve the following two problems in the ring $H(D)$ of holomorphic functions in D:

1. Given any $f \in H(D) \cap C^1(\bar{D})$ such that f is in the ideal generated by $f_1, \ldots, f_m$ in $H(D)$, how can one find a "division formula" for f, that is, find explicitly $u_1, \ldots, u_m \in H(D)$ such that $f = u_1 f_1 + \ldots + u_m f_m$ in D?

2. Can one write down a division formula with a remainder term ? That is, can one make explicit operators $R_1, \ldots, R_m, R_{m+1}$ from $H(D) \cap C^1(\bar{D})$ into $H(D)$ satisfying

$$(i) \ \forall f \in H(D) \cap C^1(\bar{D}), \ f = \sum_{j=1}^{m} (R_j f) f_j + R_{m+1} f \text{ in } D$$

$$(ii) \ \forall v_1, \ldots v_m \in H(D) \cap C^1(\bar{D}), \ R_{m+1}[v_1 f_1 + \ldots + v_m f_m] = 0 ?$$

Both problems can be easily solved when $m = 1$. Using Cauchy's formula, one can write, for $f \in H(D) \cap C^1(\bar{D})$ and $z \in D$

$$
\begin{aligned}
f(z) &= \frac{1}{2\pi i} \int_{\partial D} \frac{f(\zeta)}{\zeta - z} d\zeta \\
&= \frac{1}{2\pi i} \int_{\partial D} \frac{f(\zeta) f_1(\zeta)}{(\zeta - z) f_1(\zeta)} d\zeta = f_1(z) R_1 f(z) + R_2 f(z), \qquad (1.18)
\end{aligned}
$$

where

$$R_1 f(z) = \frac{1}{2\pi i} \int_{\partial D} \frac{f(\zeta)}{(\zeta - z)f_1(\zeta)} d\zeta$$

$$R_2 f(z) = \frac{1}{2\pi i} \int_{\partial D} \frac{f(\zeta)(f_1(\zeta) - f_1(z))}{(\zeta - z)f_1(\zeta)} d\zeta.$$

One should note here that $R_2 f(z)$ can be expressed as

$$< \bar{\partial} \frac{1}{f_1}(\zeta), \phi f \frac{f_1(\zeta) - f_1(z)}{\zeta - z} d\zeta >;$$

here ϕ is an element of $\mathcal{D}(D)$, identically equal to 1 near $\{f_1 = 0\}$. One should also point out that this procedure is far from being unique; for example, for any nonvanishing function θ in $H(D) \cap C^1(\bar{D})$, then (1.18) holds with the new operators

$$R_1^\theta f(z) = \frac{\theta(z)}{2\pi i} \int_{\partial D} \frac{f(\zeta)}{(\zeta - z)f_1(\zeta)\theta(\zeta)} d\zeta$$

$$R_2^\theta f(z) = \frac{\theta(z)}{2\pi i} \int_{\partial D} \frac{f(\zeta)(\theta(\zeta)f_1(\zeta) - \theta(z)f_1(z))}{(\zeta - z)f_1(\zeta)\theta(\zeta)} d\zeta.$$

When $m > 1$, solving problems 1 and 2 becomes more difficult. Nevertheless, there is still a way to do it, based on the formula (1.17). Let us develop here this idea for m functions, holomorphic in a neighborhood of the unit disk. This will give us the opportunity to introduce our general techniques to handle division problems. We use (1.17) with $M = 2$, that is, with two pairs $(q_1, G_1), (q_2, G_2)$, which we make explicit below.

The first pair depends on the geometry of the domain in which we try to write our division procedure, in this case, the unit disk. D is a strictly convex domain defined by the inequality $\rho < 0$, where $\rho(\zeta) = |\zeta|^2 - 1$. Because of the strict convexity of D,

$$\zeta \mapsto \rho(\zeta) + \frac{\partial \rho}{\partial \zeta}(\zeta)(z - \zeta) = \bar{\zeta} z - 1$$

cannot vanish in $\bar{D}$ provided $z \in D$. For $\epsilon > 0$, let us define, for $z \in D$ and $\zeta \in \bar{D}$,

$$q_1(z, \zeta, \epsilon) = q_1(\zeta, \epsilon) = \frac{1}{(\rho(\zeta) - \epsilon)} \frac{\partial \rho}{\partial \zeta}(\zeta) = \frac{\bar{\zeta}}{|\zeta|^2 - 1 - \epsilon}.$$

Recall that

$$\Phi_1(z, \zeta, \epsilon) = 1 + q_1(z, \zeta, \epsilon)(z - \zeta) = \frac{\bar{\zeta} z - 1}{|\zeta|^2 - 1 - \epsilon},$$

which does not vanish in $D \times \bar{D}$. Hence, we are allowed to take $G_1(t) = t^{-1}$.

The second pair is related to the generators $f_1, \ldots, f_m$. It depends also on a complex parameter λ, which we assume first to satisfy $\operatorname{Re} \lambda > 2$. The function q_2 is defined by

$$q_2(\lambda, z, \zeta) = \|f(\zeta)\|^{2(\lambda-1)} \sum_{j=1}^m \bar{f}_j(\zeta) \frac{f_j(z) - f_j(\zeta)}{z - \zeta},$$

with $\|f\|^2 = |f_1|^2 + \ldots + |f_m|^2$. We choose $G_2(t) = t^2$ and a simple computation leads to

$$\Phi_2(\lambda, z, \zeta) = 1 - \|f\|^{2\lambda} + \|f\|^{2(\lambda-1)} \sum_{j=1}^m \bar{f}_j f_j(z).$$

As we have aleady said, we shall omit ζ from the notation whenever possible, so that in the last formula we really have $\|f\| \equiv \|f(\zeta)\|$ and $\bar{f}_j \equiv \overline{f_j(\zeta)}$.

We write (1.17) for $z \in D$ for a function $f \in H(D) \cap C^1(\bar{D})$ and remark that it makes sense, provided $\operatorname{Re} \lambda > 2$, to set $\epsilon = 0$, which leads to

$$f(z) = \frac{1}{2\pi i} \int_D f \left(1 - \|f\|^{2\lambda} + \|f\|^{2(\lambda-1)} \sum_{j=1}^m \bar{f}_j f_j(z)\right)^2 \frac{d\bar{\zeta} \wedge d\zeta}{(\bar{\zeta}z - 1)^2} +$$

$$+ \frac{1}{\pi i} \int_D f \left(1 - \|f\|^{2\lambda} + \|f\|^{2(\lambda-1)} \sum_{j=1}^m \bar{f}_j f_j(z)\right) \frac{|\zeta|^2 - 1}{(\bar{\zeta}z - 1)} \bar{\partial}_\zeta q_2(\lambda, z, \zeta) \wedge d\zeta.$$

$$(1.19)$$

Our method, which consists in getting rid of λ, will be used extensively later on. The idea is to consider the right-hand side of (1.19) as a meromorphic function of λ in the whole complex plane and to compare the Laurent expansions, at $\lambda = 0$, of both sides of (1.19). To justify this trick, we need the following lemma .

Lemma 1.11. *Let* $f_1, \ldots, f_m \in H(\bar{D})$ *and let* ω *be a neighborhood of* $D \cap \{f_1 = \ldots = f_m = 0\}$ *such that* $\bar{\omega} \subset D$. *Let* $\phi \in L^1(D) \cap C^\infty(\omega)$. *Then,*

a) The mapping $\lambda \mapsto \mathcal{J}(\lambda, \phi) := \int_D \|f\|^{2\lambda} \phi d\bar{\zeta} \wedge d\zeta$ *is holomorphic in* $\operatorname{Re} \lambda > 0$ *and can be analytically continued to the whole complex plane as a meromorphic function with poles at rational points.*

b) The set of poles of $\lambda \mapsto \mathcal{J}(\lambda, \phi)$ *is included in a set* $E \subset \mathbb{Q}^-$, *the negative rationals, which is independent of* ϕ.

c) The Laurent expansion of $\lambda \mapsto \mathcal{J}(\lambda, \phi)$ *near any point* $e \in E$ *can be written as*

$$\mathcal{J}(\lambda, \phi) = \sum_{k \in \mathbb{Z}} T_{e,k}(\phi)(\lambda - e)^k,$$

where the $T_{e,k}$, $k \in \mathbb{Z}$, *are distributions in* D *of the form* $T_{e,k} = \sigma_{e,k} + \tau_{e,k}$, *with* $\sigma_{e,k} \in C^\infty$, $\operatorname{supp}(\tau_{e,k}) \subset \omega$, *and* $\operatorname{supp}(\tau_{e,k}) \subset \{f_1 = \ldots = f_m = 0\}$ *if* $k < 0$.

Proof. We use a finite covering of $\bar{D}$ by disks centered at distinct points $\alpha \in D$ and a corresponding partition of unity of $\bar{D}$, (ϕ_α), such that for any common zero of $f_1, \ldots, f_m$ in D there is a disk of the covering, which is centered at this zero, included in ω, and such that there are no other zeros of the product $f_1 \ldots f_m$ in this disk. Let us denote by $\mu(\alpha)$ the minimum of the multiplicities of α as a zero of $f_1, \ldots, f_m$. One can write

$$\mathcal{J}(\lambda, \phi) = \sum_{\alpha \in \{f_1 = \ldots = f_m = 0\}} \mathcal{J}(\lambda, \phi\phi_\alpha) + \mathcal{J}_2(\lambda, \phi). \tag{1.20}$$

$$\mathcal{J}_2(\lambda, \phi) := \sum_{\alpha \notin \{f_1 = \ldots = f_m = 0\}} \mathcal{J}(\lambda, \phi\phi_\alpha). \tag{1.21}$$

If $\alpha \notin \{f_1 = \ldots = f_m = 0\}$, then

$$\mathcal{J}(\lambda, \phi\phi_\alpha) = \int_{\mathrm{supp}(\phi_\alpha) \cap D} \exp(\lambda \log \|f\|^2) \phi\phi_\alpha d\bar{\zeta} \wedge d\zeta$$

that is,

$$\mathcal{J}(\lambda, \phi\phi_\alpha) = \int_{\mathrm{supp}(\phi_\alpha) \cap D} \left(\sum_{k=0}^\infty \lambda^k (\log \|f\|^2)^k \phi\phi_\alpha \right) d\bar{\zeta} \wedge d\zeta,$$

which is an entire function of λ, so that $\mathcal{J}_2$ is an entire function. If $\alpha \in \{f_1 = \ldots = f_m = 0\}$, we can write on $\mathrm{supp}(\phi_\alpha)$, $\|f\|^2 = |\zeta|^{2\mu(\alpha)} v_\alpha(\zeta)$, where v_α is a nonvanishing positive smooth function. Thus, for $\mathrm{Re}\,\lambda \gg 0$ and $\mu = \mu(\alpha)$,

$$\mathcal{J}(\lambda, \phi\phi_\alpha) = \frac{1}{(\mu\lambda + 1) \cdots (\mu\lambda + \mu)} \int_D |\zeta - \alpha|^{2(\lambda+1)\mu} [\frac{\partial^2}{\partial\zeta\partial\bar{\zeta}}]^\mu [\phi\phi_\alpha v_\alpha^\lambda] d\bar{\zeta} \wedge d\zeta. \tag{1.22}$$

Iterating the formula (1.22), we obtain the analytic continuation of $\mathcal{J}(\lambda, \phi\phi_\alpha)$ as a function of λ to the whole complex plane as a meromorphic function and with poles only at points of $\mathbb{Q}^-$. Parts b) and c) in the statement of the Lemma 1.11 are now consequences of the formula (1.22). $\qquad\square$

Remark 1.12. For any $j, k \in \{1, \ldots, m\}$, the function $J_{j,k}(\lambda, \phi)$ defined by

$$J_{j,k}(\lambda, \phi) = \lambda \int \|f\|^{2(\lambda-2)} \bar{f_j} f_k \bar{\partial} \|f\|^2 \wedge d\zeta$$

can be analytically continued as a holomorphic function of λ to a neighborhood of the closed half-plane $\mathrm{Re}\,\lambda \geq 0$ and takes the value 0 at $\lambda = 0$.

To prove the remark, we may replace ϕ by $\phi\phi_\alpha$ and write $\|f\|^2 = |\zeta - \alpha|^{2\mu} v_\alpha$, that is,

$$\bar{\partial}\|f\|^2 = |\zeta - \alpha|^{2\mu} \bar{\partial} v_\alpha + v_\alpha \bar{\partial}(|\zeta - \alpha|^{2\mu}).$$

Thus, we have

$$\lambda\|f\|^{2(\lambda-1)}\bar{\partial}\|f\|^2 = \bar{\partial}\|f\|^2 = \lambda v_\alpha^\lambda |\zeta - \alpha|^{2\mu\lambda} \frac{\overline{\partial(\zeta - \alpha)^\mu}}{(\zeta - \alpha)^\mu} + |\zeta - \alpha|^{2\mu\lambda}\bar{\partial}v_\alpha^\lambda.$$

The Remark 1.12 follows now from the fact that $\bar{f}_j f_k/\|f\|^2$ is bounded and that $1/(\bar{\zeta} - \bar{\alpha})$ is locally integrable in $\mathbb{C}$.

We use Lemma 1.11 and Remark 1.12 to get from (1.19) the following division formula with remainder term.

$$f(z) = \frac{1}{2\pi i} < \chi_D T_{-2,0}, f\left(\sum_{j=1}^m \bar{f}_j f_j(z)/(\bar{\zeta}z - 1)\right)^2 d\bar{\zeta} \wedge d\zeta >$$

$$+ \frac{1}{\pi i} < \chi_D T_{-2,0}, f\left(\sum_{j,k=1}^m f_j(z)\bar{f}_j\overline{f_k'}\frac{f_k(z) - f_k(\zeta)}{z - \zeta}\right)\left(\frac{1 - |\zeta|^2}{1 - \bar{\zeta}z}\right)d\bar{\zeta} \wedge d\zeta >$$

$$- \frac{1}{\pi i} < \chi_D T_{-3,0}, f\left(\sum_{j,k=1}^m f_j(z)\bar{f}_j\bar{f}_k\frac{f_k(z) - f_k(\zeta)}{z - \zeta}\right)\left(\frac{1 - |\zeta|^2}{1 - \bar{\zeta}z}\right)\bar{\partial}\|f\|^2 \wedge d\zeta >$$

$$+ \frac{1}{2\pi i} < \chi_D T_{-3,-1}, f\left(\sum_{j,k=1}^m f_j(z)\bar{f}_j\bar{f}_k\frac{f_k(z) - f_k(\zeta)}{z - \zeta}\right)\left(\frac{1 - |\zeta|^2}{1 - \bar{\zeta}z}\right)\bar{\partial}\|f\|^2 \wedge d\zeta >$$

$$+ R_{m+1}f(z), \tag{1.23}$$

where

$$R_{m+1}f(z) = \frac{1}{2\pi i} < \chi_D T_{-2,-1}, f\left(\sum_{k=1}^m \bar{f}_k \frac{f_k(z) - f_k(\zeta)}{z - \zeta}\right)\left(\frac{1 - |\zeta|^2}{1 - \bar{\zeta}z}\right)\bar{\partial}\|f\|^2 \wedge d\zeta > \tag{1.24}$$

It is clear from the Remark 1.12 that R_{m+1} satisfies the condition (ii) of Problem 2 at the beginning of this section. Therefore (1.23) is a division formula in D with remainder term.

§5 Some applications of Pompeiu's formulas. Global results

We plan to give in this section some applications of (1.17) to global division problems in the algebra of polynomials (Examples 1 and 2 below) and in the Paley-Wiener algebra of Fourier transforms of distributions with compact support (Example 3.)

Example 1. This is an analytic version of the Bezout identity in $\mathbb{C}[X]$. We consider m polynomials $P_1, \ldots, P_m$ of one complex variable, at least one of them is not constant, and the set of common zeros $\{P_1 = \cdots = P_m = 0\} = \emptyset$. We can use formula (1.17) to derive an explicit Bezout identity $1 = A_1 P_1 + \ldots + A_m P_m$ with $A_j \in \mathbb{C}[X]$. We proceed as follows, define

$$q(z, \zeta) = q(\zeta) = \frac{1}{\|P(\zeta)\|^2} \sum_{j=1}^{m} \overline{P_j(\zeta)} \frac{P_j(z) - P_j(\zeta)}{z - \zeta},$$

where, as before, $\|P\|^2 = |P_1|^2 + \ldots + |P_m|^2$. Applying (1.17) with $f = 1$, $M = 1$, $q_1 = q$, and $G(t) = t^2$ in the disk $D(0, R)$, we obtain, for $z \in D(0, R)$,

$$1 = \frac{1}{2\pi i} \int_{|\zeta| = R} \left(\sum_{j=1}^{m} \frac{\bar{P}_j P_j(z)}{\|P\|^2} \right)^2 \frac{d\zeta}{\zeta - z}$$

$$+ \frac{1}{\pi i} \int_{|\zeta| \leq R} \left(\sum_{j=1}^{m} \frac{\bar{P}_j P_j(z)}{\|P\|^2} \right) \frac{\partial q}{\partial \bar{\zeta}}(z, \zeta) d\bar{\zeta} \wedge d\zeta. \qquad (1.25)$$

As

$$\sum_{j=1}^{m} \frac{\overline{P_j(\zeta)} P_j(z)}{\|P(\zeta)\|^2} = O_z(|\zeta|^{-2}) \quad \text{when } \zeta \to \infty,$$

we have for any $z \in \mathbb{C}$ that, when $R \to \infty$, formula (1.25) becomes

$$1 = \frac{1}{\pi i} \int \left(\sum_{j=1}^{m} \frac{\bar{P}_j P_j(z)}{\|P\|^2} \right) \frac{\partial q}{\partial \bar{\zeta}}(z, \zeta) d\bar{\zeta} \wedge d\zeta = \sum_{j=1}^{m} A_j(z) P_j(z), \qquad (1.26)$$

for polynomials A_j defined by the intermediate part of the identity (1.26). Therefore, (1.26) is an explicit Bezout identity in $\mathbb{C}[X]$. It is immediate that $\deg(A_j) \leq \deg_z(\partial q(z, \zeta)/\partial \bar{\zeta})$, thus $\deg(A_j) \leq D - 1$, where $D := \max_k \deg(P_k)$. The problem we have here is that this identity does not take into account the field in which lie the coefficients of $P_1, \ldots, P_m$. This is the reason for the following Example 2, which provides also a Bezout identity such that $A_1, \ldots, A_m$ have their coefficients in the algebraic closure of the extension of $\mathbb{Q}$ by all the coefficients of $P_1, \ldots, P_m$.

Example 2. We make the same assumptions as in the previous example, except that, for the sake of definiteness, we assume that $\deg(P_1) > 0$. We apply formula (1.17) in $D(0, R)$ with $M = 2$, $f = 1$, and we let $R \gg 0$ so that all the zeroes of P_1 are in $D(0, R)$. Define two pairs of weights by

$$q_1(z, \zeta) := \frac{|P_1(\zeta)|^{2\lambda}}{P_1(\zeta)} \frac{P_1(z) - P_1(\zeta)}{z - \zeta}, \quad G_1(t) = t$$

$$q_2(z, \zeta) := \|P(\zeta)\|^{-2} \sum_{j=1}^{m} \overline{P_j(\zeta)} \frac{P_j(z) - P_j(\zeta)}{z - \zeta}, \quad G_2(t) = t \qquad (1.27)$$

Let $\operatorname{Re} \lambda \gg 0$, $|z| < R$, and $g_j(z, \zeta) := \frac{P_j(z) - P_j(\zeta)}{z - \zeta}$, $j = 1, \ldots, m$. Then, from (1.17) we obtain the identity

$$
\begin{aligned}
1 = & \frac{P_1(z)}{2\pi i} \int_{|\zeta|=R} \frac{|P_1|^{2\lambda}}{P_1} \left(\sum_{j=1}^{m} \frac{\bar{P}_j P_j(z)}{\|P\|^2} \right) \frac{d\zeta}{\zeta - z} \\
& + \frac{\lambda}{2\pi i} \int_{|\zeta| \leq R} \sum_{j=1}^{m} \frac{\bar{P}_j P_j(z)}{\|P\|^2} |P_1|^{2(\lambda-1)} g_1(z, \zeta) \overline{\partial P_1} \wedge d\zeta \\
& + \frac{P_1(z)}{2\pi i} \int_{|\zeta| \leq R} |P_1|^{2(\lambda-1)} \bar{P}_1 \frac{\partial}{\partial \bar\zeta} \left(\sum_{j=1}^{m} \frac{\bar{P}_j g_j(z, \zeta)}{\|P\|^2} \right) d\bar\zeta \wedge d\zeta \\
& + \frac{1}{2\pi i} \int_{|\zeta| \leq R} (1 - |P_1|^{2\lambda}) \left(\sum_{j=1}^{m} \frac{\bar{P}_j P_j(z)}{\|P\|^2} \right) \frac{d\zeta}{\zeta - z} + \\
& + \frac{1}{2\pi i} \int_{|\zeta| \leq R} (1 - |P_1|^{2\lambda}) \frac{\partial}{\partial \bar\zeta} \left(\sum_{j=1}^{m} \frac{\bar{P}_j g_j(z, \zeta)}{\|P\|^2} \right) d\bar\zeta \wedge d\zeta. \quad (1.28)
\end{aligned}
$$

The third term in (1.28) can be transformed via Stokes's theorem, so that

$$
\begin{aligned}
& \frac{1}{2\pi i} \int_{|\zeta| \leq R} |P_1|^{2(\lambda-1)} \bar{P}_1 \frac{\partial}{\partial \bar\zeta} \left(\sum_{j=1}^{m} \frac{\bar{P}_j g_j(z, \zeta)}{\|P\|^2} \right) d\bar\zeta \wedge d\zeta \\
= & \frac{1}{2\pi i} \int_{|\zeta|=R} |P_1|^{2(\lambda-1)} \bar{P}_1 \left(\sum_{j=1}^{m} \frac{\bar{P}_j g_j(z, \zeta)}{\|P\|^2} \right) d\zeta \\
& - \frac{\lambda}{2\pi i} \int_{|\zeta| \leq R} |P_1|^{2(\lambda-1)} \left(\sum_{j=1}^{m} \frac{\bar{P}_j g_j(z, \zeta)}{\|P\|^2} \right) \overline{\partial P_1} \wedge d\zeta .
\end{aligned}
$$

If we now let λ tend to 0, that is, we use the analytic continuation of the right-hand side of (1.28) as described in the Lemma 1.11, we obtain

$$
\begin{aligned}
1 = & \frac{P_1(z)}{2\pi i} \left(\int_{|\zeta|=R} \frac{1}{P_1} \left(\sum_{j=1}^{m} \frac{\bar{P}_j P_j(z)}{\|P\|^2} \right) \frac{d\zeta}{\zeta - z} + \int_{|\zeta|=R} \frac{1}{P_1} \left(\sum_{j=1}^{m} \frac{\bar{P}_j g_j(z, \zeta)}{\|P\|^2} \right) d\zeta \right) \\
& + < \bar\partial \frac{1}{P_1}(\zeta), \sum_{j=1}^{m} \frac{\bar{P}_j [P_j(z) g_1(z, \zeta) - P_1(z) g_j(z, \zeta)]}{\|P\|^2} d\zeta > . \quad (1.29)
\end{aligned}
$$

Let $R \to \infty$ in (1.29) to deduce the following Bezout identity

$$1 = < \bar\partial \frac{1}{P_1}(\zeta), \sum_{j=1}^{m} \frac{\bar P_j[P_j(z)g_1(z,\zeta) - P_1(z)g_j(z,\zeta)]}{\|P\|^2} d\zeta > = \sum_{j=1}^{m} A_j(z)P_j(z).$$

(1.30)

From the explicit computation of residues given by formula (1.4), we see that the coefficients of $A_1, \ldots, A_m$ belong to the algebraic closure of the extension of $\mathbb{Q}$ by the coefficients of the original polynomials P_j.

Example 3. The Fourier transform converts convolution problems in $\mathcal{C}^\infty(\mathbb{R}^n)$ or $\mathcal{C}^\infty(\Omega)$, for a convex open $\Omega \subset \mathbb{R}^n$, to division problems in an algebra of entire functions with growth conditions. Let us recall that the Fourier transform is a linear topological isomorphism between the space of compactly supported distributions in $\mathbb{R}^n$ and the space PW_n of entire functions of n complex variables satisfying

$$\exists A > 0, \ |F(\zeta)| \le A(1 + \|\zeta\|)^A e^{A\|Im\zeta\|}. \tag{1.31}$$

The Fourier transform of a radial integrable function $\sigma(\|x\|)$, of compact support, is an even function f of $\xi := \sqrt{\zeta_1^2 + \cdots + \zeta_n^2}$, where $f \in PW_1$. The function f is the Fourier-Bessel transform of the one variable function σ ([7], p.124).

We prove here as an exercise on the previous constructions, the Global Two Disks Theorem (see [2], [4] and [8] for related results.) We recall that the ball in $\mathbb{R}^n$ of center x and radius r is denoted $B(x,r)$. (In the complex plane we have been using in this section the notation $D(x,r)$.)

Theorem 1.13. *Let r_1, r_2 be two strictly positive numbers such that r_1/r_2 is not a quotient of positive zeroes of the Bessel function $J_{n/2}$. Let φ be an element in $L^1_{loc}(\mathbb{R}^n)$ such that*

$$\forall x \in \mathbb{R}^n, \quad \int_{B(x,r_1)} f(y)dy = \int_{B(x,r_2)} f(y)dy = 0. \tag{1.32}$$

Then, $f = 0$ a.e.

Proof. We may assume that $\varphi \in \mathcal{C}^\infty(\mathbb{R}^n)$, since (1.32) is clearly equivalent to the following system of convolution equations

$$\mu_j * \varphi = 0, \ j = 1, 2 \ \text{where} \ \mu_j := \chi_{B(0,r_j)}, \ j = 1, 2. \tag{1.33}$$

It is clear that this system is also satisfied by any regularization of φ. This justifies the reduction to the case where φ is smooth.

The Fourier transform of μ_j is the entire function of $\zeta \in \mathbb{C}^n$

$$\widehat{\mu_j}(\zeta) = (2\pi)^{n/2} r_j^n j_{n/2}(r_j\xi), \ \text{where} \ j_{n/2}(\xi) = \frac{J_{n/2}(\xi)}{\xi^{n/2}}.$$

Let us recall the following elementary lemma.

Lemma 1.14. *There are constants $\kappa > 0, K > 0$ so that, for any integer $l > K$, there exists $\rho_l \in]l, l+1[$ with the property*

$$\text{if } |\xi| = \rho_l \text{ or } |Im\,\xi| \geq 1 \text{ then } |j_{n/2}(r_1\xi)j_{n/2}(r_2\xi)| \geq \kappa \frac{\exp((r_1 + r_2)|Im\,\xi|))}{|\xi|^{n+1}}.$$
$$(1.34)$$

Proof of Lemma 1.14. We use here the well known asymptotic development of the Bessel function $J_{n/2}$,

$$|J_{n/2}(\xi) - \sqrt{\frac{2}{\pi\xi}}\cos(\xi - \frac{\pi(n+1)}{4})| \leq \left(\frac{3(n^2-1)}{8}\right)\sqrt{\frac{\pi}{2}}\frac{e^{|Im\,\xi|}}{|\xi|^{3/2}}\exp\left(\frac{\pi(n^2-1)}{8|\xi|}\right)$$

valid in the sector $|Arg\,\xi| \leq \theta_0 < \pi$ (see, for example, [6].) For $|\xi| \geq \pi(n^2-1)/8$, using the fact that $j_{n/2}$ is even, we obtain from the previous estimate that

$$|j_{n/2}(\xi) - \sqrt{2/\pi}|\xi|^{-\frac{n+1}{2}}\cos(\xi - \pi(n+1)/4)| \leq \frac{3e}{8}\sqrt{\pi/2}(n^2-1)\frac{e^{|Im\,\xi|}}{|\xi|^{n+3/2}}. \quad (1.35)$$

On the other hand, if we let $V = \{(2l+1)\frac{\pi}{2}, l \in \mathbb{Z}\}$ and define $d(\xi, V) = \min\{1, dist(\xi, V)\}$, the cosinus satisfies the following Lojasiewicz inequality

$$|\cos\xi| \geq \frac{1}{\pi e}d(\xi, V)\exp(|Im\,\xi|). \quad (1.36)$$

Moreover, for any $l \geq 1$, the product $\cos(r_1\xi - \pi(n+1)/4)\cos(r_2\xi - \pi(n+1)/4)$ has at most $(r_1+r_2)/\pi$ zeroes in $]l, l+1[$. It follows from the Dirichlet pigeonhole principle that there exists $\rho_l \in]l, l+1[$ such that on the circle $|\xi| = \rho_l$ we have

$$|\cos(r_1\xi - \pi(n+1)/4)\cos(r_2\xi - \pi(n+1)/4)| \geq \left(\frac{\min\{1, r_1, r_2\}}{2e(\pi + r_1 + r_2)}\right)^2 e^{(r_1+r_2)|Im\,\xi|}.$$
$$(1.37)$$

The conclusion of the lemma follows at once from (1.35) and (1.37), for a convenient choice of $K \gg 0$. $\qquad\square$

Let us return to the proof of Theorem 1.13. We show first an analytic Bezout identity in the algebra of Fourier-Bessel transforms of radial compactly supported distributions in $\mathbb{R}^n$. We use for that another variant of the Pompeiu formula with weights (1.17). Since all the elements involved here are holomorphic functions, we get this identity from the usual residue theorem in one variable. If $|z| < \rho_l$, $l \geq K$, we have for any positive integer k

$$1 = \frac{1}{2\pi i}\int\limits_{|\xi|=\rho_l}\frac{d\xi}{\xi - z}$$

$$= \frac{1}{2\pi i} \int_{|\xi|=\rho_l} \frac{\xi^k j_{n/2}(r_1\xi)j_{n/2}(r_2\xi) - z^k j_{n/2}(r_1 z)j_{n/2}(r_2 z)}{\xi^k(\xi - z)j_{n/2}(r_1\xi)j_{n/2}(r_2\xi)} d\xi$$

$$+ z^k j_{n/2}(r_1 z)j_{n/2}(r_2 z) \frac{1}{2\pi i} \int_{|\xi|=\rho_l} \frac{d\xi}{\xi^k(\xi - z)j_{n/2}(r_1\xi)j_{n/2}(r_2\xi)}$$

$$= H_l(z) + z^k j_{n/2}(r_1 z)j_{n/2}(r_2 z) A_l(z), \tag{1.38}$$

where the functions $H_l(z)$ and $A_l(z)$ are defined for $|z| \neq \rho_l$ by the integrals in (1.38). Clearly, both are holomorphic in this open set, but H_l is, in fact, entire. Moreover, as a consequence of the Residue Theorem, if $|z| > \rho_l$, we have the additional relation

$$0 = H_l(z) + z^k j_{n/2}(r_1 z)j_{n/2}(r_2 z) A_l(z). \tag{1.39}$$

Furthermore, the function H_l can be easily computed explicitly since the functions $j_{n/2}(r_1\xi)$ and $j_{n/2}(r_2\xi)$ have no common zeroes. Namely, from the calculus of residues,

$$H_l(z) = z^k j_{n/2}(r_1 z) \sum_{\substack{j_{n/2}(r_2\beta)=0 \\ |\beta|<\rho_l}} \frac{1}{r_1 \beta^k j_{n/2}(r_1\beta)j'_{n/2}(r_2\beta)} \frac{j_{n/2}(r_2 z)}{z - \beta}$$

$$+ z^k j_{n/2}(r_2 z) \sum_{\substack{j_{n/2}(r_1\alpha)=0 \\ |\alpha|<\rho_l}} \frac{1}{r_2 \alpha^k j_{n/2}(r_2\alpha)j'_{n/2}(r_1\alpha)} \frac{j_{n/2}(r_1 z)}{z - \alpha}$$

$$+ j_{n/2}(r_1 z)j_{n/2}(r_2 z) \operatorname{Res}_{\xi=0}\left(\frac{z^{k-1} + \xi z^{k-2} + \cdots + \xi^{k-1}}{\xi^k j_{n/2}(r_1\xi)j_{n/2}(r_2\xi)} \right). \tag{1.40}$$

Choosing $k = n + 1$, from (1.38), (1.39), and the estimate (1.34), we conclude there is a constant $C = C(r_1, r_2) > 0$ such that, if $|z| \neq \rho_l$, we have the two inequalities

(a) $|z^{n+1} j_{n/2}(r_1 z)j_{n/2}(r_2 z) A_l(z)| \leq C \dfrac{|z|^2 e^{(r_1+r_2)|\operatorname{Im} z|}}{|\rho_l - |z||}$,   for $|z| > \rho_l$

(b) $|z^{n+1} j_{n/2}(r_1 z)j_{n/2}(r_2 z) A_l(z)| \leq C \dfrac{|z|^2 e^{(r_1+r_2)|\operatorname{Im} z|}}{|\rho_l - |z||} + 1$, for $|z| < \rho_l$.

$$\tag{1.41}$$

From (1.41) and the Maximum Principle we deduce the existence of a positive constant $C' = C'(r_1, r_2)$ so that

$$|H_l(z)| \leq C'(|z|^2 + 1)e^{(r_1+r_2)|z|}, \quad z \in \mathbb{C}. \tag{1.42}$$

(See [2], p. 276-277, for this estimate and the following ones.) A careful estimate of $H_l(x)$, based on the identity $H_l(x) = H_l(x) - H_{l+3}(x) + H_{l+3}(x)$ for x in $[l-1, l+1]$ leads to the inequality

$$|H_l(x)| \leq C''(|x|^2 + 1), \quad x \in \mathbb{R}. \tag{1.43}$$

The estimates (1.41), (1.42), (1.43), as well as the Phragmén-Lindelöf Principle imply, for $l \geq K$ and some constant $D > 0$,

$$|H_l(z) - 1| \leq \frac{D(|z|^3 + 1)}{\rho_l} e^{(r_1 + r_2)|\operatorname{Im} z|}.$$

We conclude that H_l is the Fourier-Bessel transform of a radial distribution σ_l, and, moreover, $\sigma_l \to \delta$ in $\mathcal{E}'(\mathbb{R}^n)$ as $l \to \infty$. On the other hand, from the formula (1.40), we see that σ_l can be represented as

$$\sigma_l = \mu_1 * \tau_l^{(1)} + \mu_2 * \tau_l^{(2)}, \quad \operatorname{supp}(\tau_l^{(1)}) \subset B(0, r_2), \ \operatorname{supp}(\tau_l^{(2)}) \subset B(0, r_1).$$

Thus, from the hypothesis (1.33) on φ, we obtain that $\sigma_l * \varphi \equiv 0$, for any $l \in \mathbb{N}$. Taking the limit when $l \to \infty$, we get $\delta * \varphi \equiv 0$, and so we obtain the desired result. $\square$

Remark 1.15. The procedure developed here can be carried out even when φ is only defined in a ball $D(0, R)$, as long as $R > r_1 + r_2$ [2], and it provides a way to recover φ from its averages $\mu_1 * \varphi$ and $\mu_2 * \varphi$.

References for Chapter 1

[1] C.A. Berenstein and B.A. Taylor, A new look at interpolation theory for entire functions of one variable, Advances in Math. 33(1979), 109–143.

[2] C.A. Berenstein, R. Gay and A. Yger, Inversion of the local Pompeiu transform, J. d'Analyse Mathématique 54(1990), 259–287.

[3] C.A. Berenstein and R. Gay, *Complex variables: an introduction*, Springer-Verlag, New York, 1991.

[4] J. Delsarte, *Lectures on Topics in Mean Periodic Functions and the Two Radius Theorem*, Tata Institute, Bombay, 1961.

[5] I.M. Gelfand and G.E. Shilov, *Generalized functions*, vol.1, 2, Academic Press, New York, 1964 and 1968.

[6] L. Gatteschi, On the zeroes of certain functions with application to Bessel functions, Indag. Math.14(1952), 224–229.

[7] L. Zalcman, Analyticity and the Pompeiu Problem, Arch Rat. Mech. Anal. 47(1972), 237–254.

[8] L. Zalcman, Offbeat integral geometry, Amer. Math. Monthly 87(1980), 161–175.

Chapter 2
Integral Formulas in Several Variables

§1 Chains and cochains, homology and cohomology

We start this chapter by recalling a few basic facts about homology theory, the reader is referred to [19] for more details. Let $\mathcal{X}$ be an n-dimensional complex analytic manifold or an m-dimensional oriented $\mathcal{C}^\infty$ manifold. The orientation of $\mathbb{C}^n$ will always be chosen so that the differential form

$$(-i)^n d\bar{z}_1 \wedge dz_1 \wedge \ldots \wedge d\bar{z}_n \wedge dz_n$$

is positive. A smooth p-simplex on $\mathcal{X}$, where $p \leq 2n$ in the first case and $p \leq m$ in the second, is a pair (Δ, g), where Δ is the standard p-simplex in $\mathbb{R}^p$ (that is, $\{(t_1, \ldots, t_p) : t_j \geq 0, \sum_{j=1}^{p} t_j \leq 1\}$) and g is a $\mathcal{C}^1$ function from $\bar{\Delta}$ to $\mathcal{X}$. The choice of an order for the coordinates $t_1, \ldots, t_p$ assigns an orientation to the simplex.

If δ is a $p-1$-dimensional face of Δ, the $p-1$-dimensional simplex $(\delta, g_{|\delta})$ is called a face of the simplex (Δ, g). Such a face can be oriented with respect to the given orientation of (Δ, g) as follows: starting with a system of coordinates t, whose order is compatible with the orientation of the simplex, we make an affine change of coordinates in $\mathbb{R}^p$, $t \mapsto t'$, with positive Jacobian, in such a way that the face δ lies in $t_1' = 0$, while Δ lies in $t_1' \leq 0$; then, the coordinates $(t_2', \ldots, t_p')$ determine the orientation of $(\delta, g_{|\delta})$ induced by the orientation of (Δ, g).

A smooth p-chain on $\mathcal{X}$ is a formal linear combination $\sigma = \sum_{k=1}^{N} m_k \sigma_k$, where $m_1, \ldots, m_N \in \mathbb{Z}$ and $\sigma_1, \ldots, \sigma_N$ are smooth oriented p-simplexes. The collection of all smooth p-chains forms an Abelian group denoted by $C_p(\mathcal{X})$. The boundary operator is the group homomorphism $\partial = \partial_p : C_p(\mathcal{X}) \longrightarrow C_{p-1}(\mathcal{X})$ induced by its value on a p-simplex $\sigma = (\Delta, g)$ given by

$$\partial(\sigma) := \sum_{\delta} (\delta, g_{|\delta}),$$

where the sum takes place over all the $p-1$-dimensional faces of σ. This operator ∂ induces the homology sequence

$$\ldots \longrightarrow C_p(\mathcal{X}) \xrightarrow{\partial} C_{p-1}(\mathcal{X}) \longrightarrow \ldots .$$

The lack of exactness of this sequence is measured by the homology groups $H_p(\mathcal{X})$, $p \geq 0$,

$$H_p(\mathcal{X}) := \frac{\{c \in C_p(\mathcal{X}) : \partial c = 0\}}{\partial\{C_{p+1}(\mathcal{X})\}} = \frac{Z_p(\mathcal{X})}{B_p(\mathcal{X})}.$$

Let $C^p(\mathcal{X})$ be the group of smooth p-forms on $\mathcal{X}$. The d operator induces the De Rham cohomology sequence

$$\ldots \xrightarrow{d} C^p(\mathcal{X}) \xrightarrow{d} C^{p+1}(\mathcal{X}) \xrightarrow{d} \ldots$$

One has a corresponding sequence of cohomology groups $H^p(\mathcal{X}), p \geq 0$

$$H^p(\mathcal{X}) = \frac{\{c \in C^p(\mathcal{X}), \; \partial c = 0\}}{d\{C^{p-1}(\mathcal{X})\}}.$$

When $\mathcal{X}$ is a complex analytic manifold, we also consider the groups $\Lambda^{p,q}(\mathcal{X})$ of (p,q)-forms, i.e., locally they are linear combinations with smooth coefficients of $d\zeta_{i_1} \wedge \ldots \wedge d\zeta_{i_p} \wedge d\bar{\zeta}_{j_1} \wedge \ldots \wedge d\bar{\zeta}_{j_q}$. There is a double sequence induced on the groups $\Lambda^{p,q}(\mathcal{X})$ by the operators ∂ and $\bar{\partial}$, usually called the Dolbeault complex, and the corresponding cohomology groups are denoted $H^{p,q}(\mathcal{X})$.

The homology groups $H_p(\mathcal{X})$ can also be defined allowing the coefficients of the formal linear combinations to be real or complex numbers, instead of being integers. If one needs to make this fact explicit, one denotes them by $H_p(\mathcal{X}, \mathbb{Z})$, $H_p(\mathcal{X}, \mathbb{R})$, $H_p(\mathcal{X}, \mathbb{C})$, respectively. Any element $\omega \in H^p(\mathcal{X})$ defines a homomorphism T_ω from $H_p(\mathcal{X}, \mathbb{C})$ into $\mathbb{C}$ by

$$T_\omega(h) := \int_\alpha \omega$$

where α is a representative of the homology class h. The integration of an element ω in $C^p(\mathcal{X})$ over the chain $c = \sum_{j=1}^N m_j \sigma_j, m_1, \ldots, m_N \in \mathbb{C}$, is defined by

$$\int_c \omega := \sum_{j=1}^N m_j \int_{\Delta_j} (g_j)^* \omega.$$

Let us recall De Rham's theorem [11],

Theorem 2.1. *The map from $H^p(\mathcal{X})$ to the dual space $(H_p(\mathcal{X}, \mathbb{C}))'$ that associates T_ω to ω is a $\mathbb{C}$-linear isomorphism.*

Our main tool in this monograph is Stokes's theorem, that, in this language, can be stated as

$$\int_{\partial c} \omega = \int_c d\omega, \quad c \in C_{p+1}(\mathcal{X}, \mathbb{C}), \; \omega \in C^p(\mathcal{X}).$$

We recall from ([15], Theorem 2.7.10 and Corollary 4.2.6) that, if U is a pseudoconvex open subset in $\mathbb{C}^n$, then

$$H^p(U) = \frac{\{\omega \in \Lambda^{p,0}(U), \partial\omega = \bar{\partial}\omega = 0\}}{\partial\{\omega \in \Lambda^{p-1,0}(U), \bar{\partial}\omega = 0\}}.$$

Therefore, when U is a pseudoconvex domain in $\mathbb{C}^n$ or, more generally, when it is a Stein manifold ([15], Definition 5.1.1 and Theorem 5.2.7.), then

$$H^p(U) = 0 \text{ for } p > \dim_\mathbb{C} U. \tag{2.1}$$

§2 Cauchy's formula for test functions

To simplify the writing of multidimensional formulas, from now on, we will use the following notations.

Notations. For any multiindex $(r_1, \ldots, r_n) \in \mathbb{N}^n$, any n-uplets ζ, λ, $\mu \in \mathbb{C}^n$, and any complex number t, we let

$$r! = r_1! \ldots r_n!, \quad |r| = r_1 + \ldots + r_n$$

$$\frac{\partial^{|r|}}{\partial \zeta^r} = \frac{\partial^{r_1 + \ldots + r_n}}{\partial \zeta_1^{r_1} \ldots \partial \zeta_n^{r_n}},$$

$$\zeta^{*r} = \zeta_1^{r_1} \ldots \zeta_n^{r_n}, \quad |\zeta|^{*\lambda} = |\zeta_1|^{\lambda_1} \cdots |\zeta_n|^{\lambda_n}, \quad \zeta^r = (\zeta_1^{r_1}, \ldots, \zeta_n^{r_n})$$

$$\lambda\mu = (\lambda_1\mu_1, \ldots, \lambda_n\mu_n) \quad \lambda \cdot \mu = < \lambda, \mu > = \lambda_1\mu_1 + \cdots \lambda_n\mu_n$$

$$\underline{t} = (t, \ldots, t), \quad t\lambda = (t\lambda_1, \ldots, t\lambda_n)$$

We also need some further notation for certain differential forms; if $f_1, \ldots, f_m$ are C^1 functions we denote

$$\partial f = \partial f_1 \wedge \cdots \wedge \partial f_m, \quad \bar{\partial} f = \bar{\partial} f_1 \wedge \cdots \wedge \bar{\partial} f_m,$$

$$\overline{\partial f} = \overline{\partial f_1} \wedge \cdots \wedge \overline{\partial f_m}.$$

In particular, note that we obtain anew the familiar notation, $d\zeta = d\zeta_1 \wedge \cdots \wedge d\zeta_n$ and $d\bar{\zeta} = d\bar{\zeta}_1 \wedge \cdots \wedge d\bar{\zeta}_n$.

In this whole section U will be an open set in $\mathbb{C}^n$, K a compact subset of U, and $f_1, \ldots, f_n$, n functions holomorphic in U such that for each $j \in \{1, \ldots, n\}$, f_j is a function of the single variable ζ_j. We prove now the following generalization of Proposition 1.2.

Proposition 2.2. *Let U, K, $f_1, \ldots, f_n$ be as above. Let $\phi \in \mathcal{D}(U)$, with support in K. Then, the function of n complex variables $\lambda = (\lambda_1, \ldots, \lambda_n)$ defined by*

$$J(\lambda, \phi) = \frac{\lambda_1 \ldots \lambda_n}{(2\pi i)^n} \int_{\mathbb{C}^n} \phi |f_1(\zeta_1)|^{2(\lambda_1 - 1)} \ldots |f_n(\zeta_n)|^{2(\lambda_n - 1)} \overline{\partial f} \wedge d\zeta$$

is holomorphic in the orthant $\{\text{Re } \lambda_j > 1, \ 1 \leq j \leq n\}$ and can be analytically continued as a holomorphic function to the orthant $\{\lambda_j > -M^{-1}, j = 1, \ldots, n\}$, where M is defined as follows: if $m_j(\alpha_j)$ denotes the multiplicity of α_j as a zero of f_j and $m(\alpha) = (m_1(\alpha_1), \ldots, m_n(\alpha_n))$, then

$$M_j := \max\{m_j(\alpha_j) : \alpha \in K, \ \exists k, \ f_k(\alpha_k) = 0\}, \quad M := \max_{j=1,\ldots,n} M_j.$$

Moreover,

$$J(\underline{0}, \phi) = (-1)^{n(n-1)/2} \sum_\alpha c_\alpha \frac{\partial^{|m(\alpha)| - n}}{\partial \zeta^{m(\alpha) - \underline{1}}} [(\zeta - \alpha)^{m(\alpha)} \frac{\phi}{f_1(\zeta_1) \ldots f_n(\zeta_n)}](\alpha).$$

$$(2.2)$$

The summation takes place over the set of points $\{\alpha \in K : f_k(\alpha_k) = 0 \;\forall k\}$ *and* $c_\alpha = 1/(m(\alpha) - 1)!$.

Proof. Let us introduce a covering of K by balls, such that in each ball $B(\alpha, \rho(\alpha))$ one can write

$$f_j(\zeta_j) = u_j(\zeta_j)(\zeta_j - \alpha_j)^{m_j(\alpha_j)}, \quad u_j(\zeta_j) \neq 0 \text{ in } B(\alpha, \rho(\alpha)).$$

Let $(\phi_\alpha)_\alpha$ be a partition of unity subordinated to this covering. Let us define $E(\alpha) := \{k : 1 \leq k \leq n, m_k(\alpha_k) > 0\}$, $\nu = \nu_\alpha$ the number of elements of $E(\alpha)$, and let $\tilde{m}_k(\alpha) := m_k(\alpha) + 1$ if $k \in E(\alpha)$, $\tilde{m}_k(\alpha) := m_k(\alpha)$ if $k \notin E(\alpha)$. Using Stokes's Theorem and the formal identities (1.7), one can write, as soon as $\mathrm{Re}\,\lambda_j \gg 0$, $j = 1, \dots, n$,

$$J(\lambda, \phi\phi_\alpha) = \frac{(-1)^\nu}{(2\pi i)^n} \Big(\prod_{j \notin E(\alpha)} \lambda_j \Big) \int \frac{|\zeta - \alpha|^{*2\lambda m}}{(\zeta - \alpha)^{*m}} \Psi(\lambda, \zeta) d\bar{\zeta} \wedge d\zeta,$$

where

$$\Psi(\lambda, \zeta) = |u|^{*2\lambda} \left(\frac{\prod\limits_{j \notin E(\alpha)} \overline{f'_j(\zeta_j)}}{\big(\prod\limits_{j \in E(\alpha)} u_j \big)\big(\prod\limits_{j \notin E(\alpha)} |u_j|^2 \big)} \right) \left(\prod_{j \in E(\alpha)} \frac{\partial}{\partial \bar\zeta_j} \right) [\phi\phi_\alpha](\zeta).$$

We integrate by parts to obtain

$$J(\lambda, \phi\phi_\alpha) = \frac{(-1)^{|\tilde{m}|}}{(2\pi i)^n P_\alpha(\lambda)} \int |\zeta - \alpha|^{*2\lambda m} \left(\frac{\bar\zeta - \bar\alpha}{\zeta - \alpha} \right)^{*m} \frac{\partial^{|m|}}{\partial\bar\zeta^m} \Psi(\lambda, \zeta) \, d\bar\zeta \wedge d\zeta,$$

where

$$P_\alpha(\lambda) = \prod_{k \in E(\alpha)} \Big(\prod_{j=1}^{m_k} (m_k \lambda_k + j) \Big).$$

Introducing polar coordinates, with center α, as it was done in Proposition 1.5, we obtain immediately that the function $\lambda \mapsto J(\lambda, \phi\phi_\alpha)$ has an analytic continuation as a holomorphic function to the orthant determined by the inequalities $2m_k \,\mathrm{Re}\,\lambda_k + 1 > -1$, $k \in E(\alpha)$. Furthermore, the value at $\lambda = \underline{0}$ is 0 if $E(\alpha) \neq \{1, \dots, n\}$. Note that, in this case, the right-hand side of (2.2) also vanishes, as one can check immediately.

Let us verify now (2.2) in the remaining case $E(\alpha) = \{1, \dots, n\}$. When $\mathrm{Re}\,\lambda_j \gg 0$, $j = 1, \dots, n$, if $m_1 > 1$, using the formal identity (1.7) and repeated integrations by part with respect to ζ_1, one obtains

$$J(\lambda, \phi\phi_\alpha) =$$
$$= \frac{(-1)^n}{(2\pi i)^n} \int \frac{|\zeta - \alpha|^{*2\lambda m}}{(\zeta - \alpha)^{*m}} \frac{|u|^{*2\lambda}}{u_1 \cdots u_n} \frac{\partial^n}{\partial\bar\zeta_1} (\phi\phi_\alpha) \, d\bar\zeta \wedge d\zeta$$

$$= \frac{(-1)^{\frac{n(n+1)}{2}}}{(2\pi i)^n} \frac{(-1)^{m_1-1}}{(\lambda_1 m_1 - 1) \cdots (m_1(\lambda_1 - 1) + 1)} \times$$

$$\times \int \frac{|\zeta_1 - \alpha_1|^{2\lambda_1 m_1}}{(\zeta_1 - \alpha_1)} \frac{|\zeta_2 - \alpha_2|^{2\lambda_2 m_2}}{(\zeta_2 - \alpha_2)^{2m_2}} \cdots \frac{|\zeta_n - \alpha_n|^{2\lambda_n m_n}}{(\zeta_n - \alpha_n)^{2m_n}} \Theta_\lambda^{(1)} \bigwedge_{j=1}^{n} (d\bar{\zeta}_j \wedge d\zeta_j),$$

$$(2.3)$$

with

$$\Theta_\lambda^{(1)} = \frac{\partial^{m_1-1}}{\partial \zeta_1^{m_1-1}} \left(\frac{\partial^n(\phi\phi_{alpha})}{\partial \bar{\zeta}^{\underline{1}}} [\frac{|u|^{*2\lambda}}{u_1 \cdots u_n}] \right)$$

The purpose of the step above was to obtain $\zeta_1 - \alpha_1$ without any powers with exponents strictly larger than 1 in the denominator, hence, if $m_1 = 1$ we skip this step. Iterating the process, we have, for Re $\lambda_j \gg 0$, $j = 1, \ldots, n$,

$$J(\lambda, \phi\phi_\alpha) = \frac{1}{(2\pi i)^n} \frac{(-1)^{|m|}}{\prod_{\substack{1 \le j \le n \\ m_j > 1}} \prod_{k=1}^{m_j-1} (\lambda_j m_j - k)} \int \frac{|\zeta - \alpha|^{*2\lambda m}}{(\zeta - \alpha)^{*m}} \Theta_\lambda^{(n)} d\bar{\zeta} \wedge d\zeta. \quad (2.4)$$

where

$$\Theta_\lambda^{(n)}(\zeta) = \frac{\partial^{|m|-n}}{\partial \zeta^{m-\underline{1}}} \left(\frac{\partial^n(\phi\phi_\alpha)}{\partial \bar{\zeta}^{\underline{1}}} [\frac{|u|^{*2\lambda}}{u_1 \cdots u_n}] \right)$$

$$= \frac{\partial^n}{\partial \bar{\zeta}^{\underline{1}}} \left(\frac{\partial^{|m|-n}}{\partial \zeta^{m-\underline{1}}} [\frac{\phi\phi_\alpha}{u_1 \cdots u_n}] \right) + \sum_{j=1}^{n} \lambda_j \tau_j(\lambda, \zeta). \quad (2.5)$$

where τ_j, $j = 1, \ldots, n$ are C^∞ functions in ζ which depend holomorphically of the complex parameters λ.

Note that the product of affine factors in λ_j in (2.4) makes appear removable singularities in $J(\lambda, \phi\phi_\alpha)$, since they lie in the orthant where we already know $J(\lambda, \phi\phi_\alpha)$ is holomorphic. The value of $J(\lambda, \phi\phi_\alpha)$ is easy to compute from (2.4) and (2.5). We need to use Pompeiu's formula in order to recover formula (2.2) and therefore conclude the proof of the proposition. □

As a corollary we have the following

Proposition 2.3. Let $\phi \in \mathcal{D}(\mathbb{C}^n)$. For any $z \in \mathbb{C}^n$ one has the following identity

$$\phi(z) = \lim_{\lambda \to \underline{0}} (-1)^{n(n-1)/2} \frac{\lambda_1 \ldots \lambda_n}{(2\pi i)^n} \int |\zeta - z|^{*2(\lambda-\underline{1})} \phi(\zeta) d\bar{\zeta} \wedge d\zeta. \quad (2.6)$$

Remark 2.4. Proposition 2.3, known as Cauchy's formula, is valid whenever ϕ is C^1 with compact support. Here the limit when λ tends to $\underline{0}$ denotes the value at the origin of the analytic continuation of the function.

Proof. We apply the Proposition 2.2 to the holomorphic affine functions $f_j(\zeta) = \zeta_j - z_j$, $j = 1, \ldots, n$. □

Formula (2.6) can be used to derive the classical Cauchy formula for holomorphic functions of several complex variables. Before proving this, let us point out the following lemma that will be useful to us later on.

Lemma 2.5. *Let U be a bounded open set in $\mathbb{C}^n$ with piecewise smooth C^1 boundary. Let $(\chi_k)_k$ be a sequence of smooth functions with compact support in U converging (in the sense of distributions) to the characteristic function of U, χ_U. Let $p, q \in \mathbb{N}$ and ϕ a C^1 $(n, n-1)$-form in $\mathbb{C}^n$, then*

$$\lim_{k \mapsto \infty} \int \chi_k^p (1 - \chi_k)^q \bar{\partial} \chi_k \wedge \phi = -\frac{p!q!}{(p+q+1)!} \int_{\partial U} \phi,$$

with the orientation on ∂U induced by the orientation of $\mathbb{C}^n$.

Proof. Since χ_k vanishes on ∂U, applying Stokes's theorem we obtain

$$\int \chi_k^p (1 - \chi_k)^q \bar{\partial} \chi_k \wedge \phi = \int \chi_k^p (\sum_{l=0}^{q} (-1)^l \binom{q}{l} \chi_k^l) \bar{\partial} \chi_k \wedge \phi$$

$$= \sum_{l=0}^{q} (-1)^l \binom{q}{l} \bar{\partial} [\frac{\chi_k^{p+l+1}}{p+l+1}] \wedge \phi$$

$$= -\sum_{l=0}^{q} (-1)^l \binom{q}{l} \int \frac{\chi_k^{p+l+1}}{p+l+1} \bar{\partial} \phi.$$

As $\chi_k^{p+l+1} \to \chi_U$ for any $l \in \mathbb{N}$, we have

$$\lim_{k \mapsto \infty} \int \chi_k^p (1 - \chi_k)^q \bar{\partial} \chi_k \wedge \phi = -(\sum_{l=0}^{q} (-1)^l \binom{q}{l} \frac{1}{p+l+1}) \int_{\partial U} \phi$$

$$= -(\int_0^1 (1-t)^q t^p dt) \int_{\partial U} \phi$$

$$= -B(q+1, p+1) \int_{\partial U} \phi.$$

This concludes the proof of Lemma 2.5. □

Let us deduce Cauchy's formula from (2.6). Suppose that h is holomorphic in a neighborhood of a polydisk $D(a_1, r_1) \times \cdots \times D(a_n, r_n)$. Let $z \in D(a_1, r_1) \times \cdots \times D(a_n, r_n)$ and consider test functions of one complex variable $\chi^{(1)}, \ldots, \chi^{(n)}$, respectively supported by the disks $D(a_j, r_j)$, and such that $\chi^{(j)} \equiv 1$ near z_j. Then, for $\mathrm{Re}\, \lambda_j \gg 0$, $j = 1, \ldots, n$,

$$\frac{\lambda_1 \ldots \lambda_n}{(2\pi i)^n} \int |\zeta - z|^{*2(\lambda - 1)} h(\zeta) \chi^{(1)}(\zeta) \ldots \chi^{(n)}(\zeta) d\bar{\zeta} \wedge d\zeta$$

$$= \frac{(-1)^n}{(2\pi i)^n} \int h(\zeta) \frac{|\zeta_1 - z_1|^{2\lambda_1} \cdots |\zeta_n - z_n|^{2\lambda_n}}{(\zeta_1 - z_1) \cdots (\zeta_n - z_n)} \bar{\partial} \chi^{(1)} \wedge \cdots \wedge \bar{\partial} \chi^{(n)} \wedge d\zeta.$$

Using Lemma 2.5, with $\chi^{(j)}$ an approximation of $\chi_{D(a_j,r_j)}$, we conclude that

$$(-1)^{\frac{n(n-1)}{2}} \frac{\lambda_1 \dots \lambda_n}{(2\pi i)^n} \int |\zeta - z|^{*2(\lambda-\underline{1})} h(\zeta) d\bar\zeta \wedge d\zeta$$

$$= \frac{1}{(2\pi i)^n} \int_{\partial D(a_1,r_1) \times \dots \times \partial D(a_n,r_n)} h(\zeta) \frac{|\zeta - z|^{*2\lambda}}{(\zeta_1 - z_1) \dots (\zeta_n - z_n)} d\zeta.$$

Remark that, in the previous formula, every one-dimensional circle $\partial D(a_j, r_j)$ has the usual counterclockwise orientation. When $\lambda = \underline{0}$, we obtain the classical Cauchy representation formula

$$h(z) = \frac{1}{(2\pi i)^n} \int_{\partial D(a_1,r_1) \times \dots \times \partial D(a_n,r_n)} \frac{h(\zeta)}{(\zeta_1 - z_1) \dots (\zeta_n - z_n)} d\zeta. \qquad (2.7)$$

This result can be stated as follows.

Proposition 2.6. *Let $D = D_1 \times \dots \times D_n$ be a polydomain in $\mathbb{C}^n$ such that each D_j is a bounded domain in $\mathbb{C}$ with piecewise smooth boundary and let $h \in H(D) \cap C^0(\bar D)$, then, for $z \in D$,*

$$h(z) = \frac{1}{(2\pi i)^n} \int_{\partial D(a_1,r_1) \times \dots \times \partial D(a_n,r_n)} \frac{h(\zeta)}{(\zeta_1 - z_1) \dots (\zeta_n - z_n)} d\zeta. \qquad (2.8)$$

We will give later another proof of (2.8), as well as an extension of it, the Cauchy-Weil formula.

§3 Weighted Bochner-Martinelli formulas

We follow in this section the construction of weighted Bochner-Martinelli formulas due to B. Berndtsson and M. Andersson [6],[7]. Our approach will allow us to obtain representation formulas for functions (and not for differential forms, as in [6]); it is essentially due to M. Andersson and M. Passare [3].

It is well known that a fundamental solution for the Laplace equation in $\mathbb{R}^{2n}$ is $c_n \|\zeta\|^{2-2n}$. From this and from the Green-Ostrogradski formula, it follows easily that one has a representation formula for functions in $C^1(\mathbb{R}^{2n})$ which extends Pompeiu's formula (1.15). Our objective in this section is to write down a whole range of such representation formulas for holomorphic functions. Our key trick will be to introduce some weights in our C^1 Cauchy's formula (2.6).

Let U be a bounded domain in $\mathbb{C}^n$ with piecewise smooth boundary, and ω a relatively compact open set such that $\bar\omega \subset U$. Let $q_1, \dots, q_M$ be M $\mathbb{C}^n$-valued functions which are in $C^1(\bar\omega \times \bar U)$, with $q_k = (q_{k,1}, \dots, q_{k,n})$, $k = 1, \dots, M$ and $G_1, \dots, G_M$, M functions of one complex variable which are such that for any

k, G_k is defined and holomorphic in a neighborhood of the image of $\bar{\omega} \times \bar{U}$ by the mapping

$$(z,\varsigma) \mapsto 1 + \sum_{j=1}^{n} q_{k,j}(z,\varsigma)(z_j - \varsigma_j) = 1+ <q_k(z,\varsigma), z-\varsigma> .$$

We moreover assume, for any k, $G_k(1) = 1$.

For reasons of convenience we denote, for $1 \le k \le M$ and $\alpha \in \mathbb{N}$,

$$\Gamma_k^{(\alpha)}(z,\varsigma) := \frac{d^\alpha}{dt^\alpha} G_k|_{t=1+<q_k(z,\varsigma),z-\varsigma>}$$

We also denote as Q_k the (1,0) form

$$Q_k = \sum_{j=1}^{n} q_{k,j}(z,\varsigma)d\varsigma_j ,$$

and set, for any strictly positive integer r,

$$\bar{\partial} Q_k^r = \bar{\partial} Q_k \wedge \ldots \wedge \bar{\partial} Q_k \quad (r-\text{times}).$$

Here, the differential operator $\bar{\partial}$ acts on the variable ς. Since, $\bar{\partial} Q_k$ is a 2-form, there is no ambiguity in the definition of the wedge product. We take here the opportunity to introduce some further notation.

Notation. Let $\Omega_1, \ldots, \Omega_q$ be a collection of differential forms (not necessarily of the same degree), we denote

$$\Omega := \bigwedge_{j=1}^{q} \Omega_j := \Omega_1 \wedge \ldots \wedge \Omega_q .$$

The main point is that the product occurs in the increasing order of the indices. Refering to same formula, if $1 \le k \le q$, we denote

$$\Omega_{[k]} := \Omega_1 \wedge \ldots \wedge \widehat{\Omega_k} \wedge \ldots \wedge \Omega_q ,$$

where the hat implies that the term Ω_k is missing.

Note that all the ingredients relative to the choice of pairs of weights have been already introduced before Proposition 1.10. The additional tool we need here is related to the division by $z - \varsigma$, which can be performed without ambiguity when $n = 1$, and needs some precautions for $n > 1$. To solve this problem, we consider an open neighborhood V of ∂U and introduce a function $s : \bar{\omega} \times \bar{V} \to \mathbb{C}^n$, which is C^1 on $\bar{\omega} \times \bar{V}$ and satisfies

$$< s(z,\varsigma), z-\varsigma > = \sum_{j=1}^{n} s_j(z,\varsigma)(z_j - \varsigma_j) \neq 0, \ z \in \omega, \ \varsigma \in \partial U. \qquad (2.9)$$

Division by $z - \zeta$ can now be performed with the help of the section s. Let χ be a test function supported in U, which is identically one outside V. We have

$$\chi(\zeta) = 1 + (1 - \chi(\zeta)) \frac{\sum\limits_{j=1}^{n} s_j(z,\zeta)(z_j - \zeta_j)}{< s(z,\zeta), \zeta - z >} = 1 + < q^{(\chi)}(z,\zeta), z - \zeta >,$$

where

$$q^{(\chi)}(z,\zeta) = (1 - \chi(\zeta)) \frac{s(z,\zeta)}{< s(z,\zeta), \zeta - z >}.$$

Note that the choice of s is far from unique and that all these constructions can be carried out when $\mathbb{C}^n$ is replaced by a Stein manifold (see for example, [17] or [10], p. 580-581.) We let S be the $(1,0)$ differential form $\sum s_j d\zeta_j$. We have the following theorem.

Theorem 2.7. *Let h be holomorphic in U, C^1 in $\bar{U}$. Let ω, $q_1, \ldots, q_M$, and $G_1, \ldots, G_M$ be as above. Then, for $z \in \omega$ we have*

$$h(z) = \frac{1}{(2\pi i)^n} \left(\int_U h(\zeta) P(z,\zeta) + \int_{\partial U} h(\zeta) K(z,\zeta) \right),$$

where

$$P(z,\zeta) = \sum_{|\alpha|=n} \frac{1}{\alpha!} \Gamma_1^{(\alpha_1)} \ldots \Gamma_M^{(\alpha_M)} \bar{\partial} Q_1^{\alpha_1} \wedge \ldots \wedge \bar{\partial} Q_M^{\alpha_M}(z,\zeta)$$

$$K(z,\zeta) = \sum_{\alpha_0+|\alpha|=n-1} \frac{1}{\alpha!} \Gamma_1^{(\alpha_1)} \ldots \Gamma_M^{(\alpha_M)} \frac{S \wedge (\bar{\partial} S)^{\alpha_0} \wedge \bar{\partial} Q_1^{\alpha_1} \wedge \ldots \wedge \bar{\partial} Q_M^{\alpha_M}}{< s(z,\zeta), \zeta - z >^{\alpha_0+1}}(z,\zeta).$$

Here are used for multiindexes $\alpha \in \mathbb{N}^M$ the notations previously defined (for $|\alpha|, \alpha!$).

Proof. Fix for the moment a function χ such that $\chi \in \mathcal{D}(U)$ and satisfies $\chi \equiv 1$ in ω and assume further that (2.9) holds for $z \in \omega$, $\zeta \in \text{supp}(1 - \chi)$. We introduce

$$q_0(z,\zeta) = q^{(\chi)}(z,\zeta) \quad G_0(t) = t^{n+1}.$$

As one can see at once, $\Gamma_0^{(0)}(z,\zeta) = (\chi(\zeta))^{n+1}$, which implies that any of the functions $\Gamma_0^{(\alpha_0)}(z,\zeta)$, $\alpha_0 \leq n$, has compact support. We use now the Proposition 2.3 for

$$\phi(\zeta) = h(\zeta) \prod_{k=0}^{M} \Gamma_k^{(0)}(z,\zeta).$$

The idea, as in the previous section, is to use Stokes's theorem to raise the exponent of $|\zeta_1 - z_1|$ in (2.6); when $\mathrm{Re}\ \lambda_j \gg 0$, $j = 1, \ldots, n$,

$$\bar{\partial}\left[\frac{|\zeta_1 - z_1|^{2\lambda_1}}{\zeta_1 - z_1}\right] = \lambda_1 |\zeta_1 - z_1|^{2(\lambda_1 - 1)} d\bar{\zeta}_1$$

$$\bar{\partial}[\lambda_2 \ldots \lambda_n |\zeta_2 - z_2|^{2(\lambda_2 - 1)} \ldots |\zeta_n - z_n|^{2(\lambda_n - 1)} d\bar{\zeta}_2 \wedge \ldots \wedge d\bar{\zeta}_n \wedge h(\zeta) \prod_{k=0}^{M} \Gamma_k^{(0)}(z, \zeta) d\zeta]$$

$$= \lambda_2 \cdots \lambda_n |\zeta_2 - z_2|^{2(\lambda_2 - 1)} \ldots |\zeta_n - z_n|^{2(\lambda_n - 1)} \times$$

$$\times \left(\sum_{k=1}^{M} \Gamma_0^{(0)} \cdots \Gamma_k^{(1)} \cdots \Gamma_M^{(0)} \left(\sum_{j=1}^{n} \frac{\partial q_{k,j}}{\partial \bar{\zeta}_1}(z_j - \zeta_j) \right) \right) h(\zeta) d\bar{\zeta} \wedge d\zeta$$

Using Stokes's formula for $\mathrm{Re}\ \lambda_1 \gg 0$ (while keeping $\mathrm{Re}\ \lambda_j$, for $2 \leq j \leq n$, sufficiently large), we obtain from (2.6)

$$h(z) = \lim_{\lambda \to \underline{0}} (-1)^{\frac{n(n+1)}{2}} \int \frac{|\zeta_1 - z_1|^{2\lambda_1}}{\zeta_1 - z_1} \left(\bigwedge_{j=2}^{n} \lambda_j |\zeta_j - z_j|^{2(\lambda_j - 1)} d\bar{\zeta}_j \right) \wedge R(z, \zeta) \wedge d\zeta$$

$$(2.10)$$

with

$$R(z, \zeta) = h(\zeta) \sum_{k=0}^{M} \Gamma_0^{(0)} \cdots \Gamma_k^{(1)} \cdots \Gamma_M^{(0)} \left(\sum_{j=1}^{n} \frac{\partial q_{k,j}}{\partial \bar{\zeta}_1}(z_j - \zeta_j) \right) d\bar{\zeta}_1.$$

Let us write

$$R(z, \zeta) = (z_1 - \zeta_1) h(\zeta) \left(\sum_{k=0}^{M} \Gamma_0^{(0)} \cdots \Gamma_k^{(1)} \cdots \Gamma_M^{(0)} \frac{\partial q_{k,1}}{\partial \bar{\zeta}_1} \right) d\bar{\zeta}_1$$

$$+ h(\zeta) \left(\sum_{j=2}^{n} (z_j - \zeta_j) \left(\sum_{k=0}^{M} \Gamma_0^{(0)} \cdots \Gamma_k^{(1)} \cdots \Gamma_M^{(0)} \frac{\partial q_{k,j}}{\partial \bar{\zeta}_1} \right) \right) d\bar{\zeta}_1$$

$$= R_0(z, \zeta) + R_1(z, \zeta).$$

Consider first

$$\mathcal{I}_1(\lambda) = \int \frac{|\zeta_1 - z_1|^{2\lambda_1}}{\zeta_1 - z_1} \left(\bigwedge_{j=2}^{n} \lambda_j |\zeta_j - z_j|^{2(\lambda_j - 1)} d\bar{\zeta}_j \right) \wedge R_1(z, \zeta) \wedge d\zeta.$$

When $\mathrm{Re}\ \lambda_1 > 0$, this is a holomorphic function of $\lambda_2, \ldots, \lambda_n$ which can be analytically continued to $\{\mathrm{Re}\ \lambda_j > 0, j = 2, \ldots, n\}$. From Proposition 2.3, the value $\mathcal{I}_1(\lambda_1, 0, \ldots, 0)$ is 0 because $R_1(z, \zeta)$ can be written

$$(\zeta_2 - z_2) R_{1,2} + \ldots + (\zeta_n - z_n) R_{1,n}.$$

Moreover, using integration by parts, one can show that the analytic continuations of

$$\lambda \mapsto \mathcal{I}_\epsilon(\lambda) = \int \frac{|\zeta_1 - z_1|^{2\lambda_1}}{\zeta_1 - z_1} \left(\bigwedge_{j=2}^{n} \lambda_j |\zeta_j - z_j|^{2(\lambda_j - 1)} d\bar{\zeta}_j \right) \wedge R_\epsilon(z, \zeta) \wedge d\zeta$$

for $\epsilon = 0, 1$, exist up to the origin. Moreover, we have $\mathcal{I}_1(\underline{0}) = 0$. Therefore, we can conclude

$$h(z) = \lim_{\lambda \to \underline{0}} \left[(-1)^{\frac{n(n-1)}{2} + n - 1} \int h(\zeta) |\zeta_1 - z_1|^{2\lambda_1} \left(\bigwedge_{j=2}^{n} \lambda_j |\zeta_j - z_j|^{2(\lambda_j - 1)} d\bar{\zeta}_j \right) \right.$$

$$\left. \wedge \left(\sum_{k=0}^{M} \Gamma_0^{(0)} \cdots \Gamma_k^{(1)} \cdots \Gamma_M^{(0)} \bar{\partial} q_{k,1} \wedge d\zeta \right) \right]$$

$$= \lim_{\lambda \to \underline{0}} \left[(-1)^{\frac{n(n-1)}{2} + n - 1} \int h(\zeta) \left(\bigwedge_{j=2}^{n} \lambda_j |\zeta_j - z_j|^{2(\lambda_j - 1)} d\bar{\zeta}_j \right) \right.$$

$$\left. \wedge \left(\sum_{k=0}^{M} \Gamma_0^{(0)} \cdots \Gamma_k^{(1)} \cdots \Gamma_M^{(0)} \bar{\partial} q_{k,1} \wedge d\zeta \right) \right]$$

It is clear that this process can be continued. As a matter of fact,

$$\bar{\partial} \left[\sum_{k=0}^{M} \Gamma_0^{(0)} \cdots \Gamma_k^{(1)} \cdots \Gamma_M^{(0)} \bar{\partial} q_{k,1} \right]$$

$$= \sum_{k,l=0}^{M} \Gamma_0^{(0)} \cdots \Gamma_l^{(1)} \cdots \Gamma_k^{(1)} \cdots \Gamma_M^{(0)} \left(\sum_{j=1}^{n} (z_j - \zeta_j) \bar{\partial} q_{l,j} \right) \wedge \bar{\partial} q_{k,1}.$$

$$\tag{2.11}$$

The coefficient of $z_1 - \zeta_1$ vanishes in (2.11) because of the anticommutativity of the wedge product. On the other hand, it can be seen that χ raised to the power $n + 1 - 1 = n$ is present as a factor in the right-hand side of (2.11), so that we have compactly supported functions in (2.11). Repeating n times this procedure, we can represent $h(z)$, for $z \in \omega$, as the integral

$$\frac{1}{(2\pi i)^n} \int h \left(\sum_{\alpha_0 + |\alpha| = n} \frac{1}{\alpha_0! \alpha!} \Gamma_0^{(\alpha_0)} \cdots \Gamma_M^{(\alpha_M)} (\bar{\partial} Q_0)^{\alpha_0} \wedge \ldots \wedge (\bar{\partial} Q_M)^{\alpha_M} (z, \zeta) \right)$$

$$\tag{2.12}$$

We also have

$$\bar{\partial} Q_0 = -\frac{\bar{\partial} \chi \wedge S}{< s, \zeta - z >} + (1 - \chi) \bar{\partial} [\frac{S}{< s, \zeta - z >}].$$

From $S \wedge S = 0$, it follows that for $\alpha_0 \in \mathbb{N}^*$,

$$(\bar{\partial}Q_0)^{\alpha_0} =$$

$$= (1-\chi)^{\alpha_0} \left(\bar{\partial}[\frac{S}{<s,\zeta-z>}] \right)^{\alpha_0} - \alpha_0 (1-\chi)^{\alpha_0-1} \left(\bar{\partial}[\frac{S}{<s,\zeta-z>}] \right)^{\alpha_0-1}$$

$$\wedge \frac{\bar{\partial}\chi \wedge S}{<s,\zeta-z>}$$

$$= (1-\chi)^{\alpha_0} \left(\bar{\partial}[\frac{S}{<s,\zeta-z>}] \right)^{\alpha_0} - \alpha_0 (1-\chi)^{\alpha_0-1} \frac{\bar{\partial}\chi \wedge S \wedge (\bar{\partial}S)^{\alpha_0-1}}{<s,\zeta-z>^{\alpha_0+1}} .$$

$$(2.13)$$

Besides, for $\alpha_0 \in \mathbb{N}$,

$$\Gamma_0^{(\alpha_0)}(z,\zeta) = \frac{(n+1)!}{(n+1-\alpha_0)!} \chi^{n+1-\alpha_0}(\zeta). \qquad (2.14)$$

Clearly $\chi^p(1-\chi)^q$ converges to zero (as distribution) when χ tends to the characteristic function of U. So we conclude from (2.12) that

$$h(z) = \frac{1}{(2\pi i)^n} \int h(\zeta)P(z,\zeta)$$

$$- \lim_{\chi \to \chi_U} \frac{1}{(2\pi i)^n} \int h(\zeta) \left(\sum_{\substack{\alpha_0+|\alpha|=n \\ \alpha_0>0}} \alpha_0 \binom{n+1}{\alpha_0} \chi^{n+1-\alpha_0}(1-\chi)^{\alpha_0-1}\bar{\partial}\chi \wedge W_\alpha(z,\zeta) \right)$$

$$(2.15)$$

where $W_\alpha := W_\alpha(z,\zeta)$ is the differential form

$$W_\alpha = \frac{1}{\alpha!}\Gamma_1^{(\alpha_1)} \cdots \Gamma_M^{(\alpha_M)} \left(\frac{S \wedge (\bar{\partial}S)^{\alpha_0-1}}{<s,\zeta-z>^{\alpha_0+1}} \right) \wedge (\bar{\partial}Q_1)^{\alpha_1} \wedge \ldots \wedge (\bar{\partial}Q_M)^{\alpha_M}(z,\zeta)$$

Now the conclusion of the theorem follows from Lemma 2.5 . The point that we need to mention is that we make a change of indexation in formula (2.15), namely α_0 becomes $\alpha_0 - 1$. □

There are some variants of this last result. To give an example of one of them, let us introduce, besides our original data U, h, ω, $q_1,\ldots,q_M$, s, and $G_1,\ldots G_M$, a $\mathbb{C}^n$-valued application τ in $\mathcal{C}^1(\bar{\omega}\times\bar{U})$. Let T be the corresponding $(1,0)$ form, $T = \sum \tau_j d\zeta_j$. We have then the following proposition.

Proposition 2.8. *Let $\Gamma_h(z,\zeta)$ be the function defined in $\omega \times \bar{U}$ by*

$$\Gamma_h(z,\zeta) = h(\zeta) + <\tau(z,\zeta),\, z-\zeta> .$$

One has, for any z in ω, the following representation formula

$$h(z) = \frac{1}{(2\pi i)^n} \left(\int_U \tilde{P}_h(z,\zeta) + \int_{\partial U} h(\zeta) K(z,\zeta) \right),$$

where the two kernels are defined by

$$\tilde{P}_h(z,\zeta) = \Gamma_h(z,\zeta) \sum_{|\alpha|=n} \frac{1}{\alpha!} \Gamma_1^{(\alpha_1)} \dots \Gamma_M^{(\alpha_M)} \bar{\partial} Q_1^{\alpha_1} \wedge \dots \wedge \bar{\partial} Q_M^{\alpha_M}(z,\zeta)$$

$$+ \sum_{|\alpha|=n-1} \frac{1}{\alpha!} \Gamma_1^{(\alpha_1)} \dots \Gamma_M^{(\alpha_M)} \bar{\partial} Q_1^{\alpha_1} \wedge \dots \wedge \bar{\partial} Q_M^{\alpha_M} \wedge \bar{\partial}_\zeta T(z,\zeta)$$

and, as before,

$$K(z,\zeta) = \sum_{\alpha_0+|\alpha|=n-1} \frac{1}{\alpha!} \Gamma_1^{(\alpha_1)} \dots \Gamma_M^{(\alpha_M)} \frac{S \wedge (\bar{\partial} S)^{\alpha_0} \wedge \bar{\partial} Q_1^{\alpha_1} \wedge \dots \wedge \bar{\partial} Q_M^{\alpha_M}}{< s(z,\zeta), \zeta - z >^{\alpha_0+1}}(z,\zeta).$$

Proof. The idea of the proof of this result is completely similar to what we did above. We apply Proposition 2.3 to the function ψ defined as

$$\psi(\zeta) = \Gamma_h(z,\zeta) \prod_{k=0}^{M} \Gamma_k^{(0)}(z,\zeta).$$

with exactly the same choice as before for the pair (q_0, G_0). This proposition will be useful for us in our division techniques later.   □

Taking $q_1 = q_2 = \dots = q_M = 0$ in Theorem 2.7 we deduce the classical Bochner-Martinelli formula, which follows.

Theorem 2.9. *Let U be a bounded domain in $\mathbb{C}^n$ with piecewise smooth boundary. Let s be a $\mathbb{C}^n$-valued function defined on a neighborhood of ∂U and of class C^1 be such that $< s(\zeta), \zeta - z_0 > \neq 0$ for any $\zeta \in \partial U$ and some $z_0 \in U$. Then, for $h \in H(U) \cap C^0(\bar{U})$*

$$h(z_0) = \frac{(n-1)!(-1)^{\frac{n(n-1)}{2}}}{(2\pi i)^n} \int_{\partial U} h(\zeta) \frac{\sum_{k=1}^{n}(-1)^{k-1} s_k(\zeta) ds_{[k]} \wedge d\zeta}{< s(\zeta), \zeta - z_0 >^n}. \qquad (2.16)$$

Proof. We perturb slightly the domain U, changing it into $\tilde{U}$, in such a way that $f \in C^1(\tilde{U})$; we then introduce a compactly supported smooth function χ such that $\text{supp}(\chi) \subset\subset U$ and that s is defined outside the set $E = \{\chi = 1\}$; moreover, we can assume that, for any ζ in U, outside E, the function $\zeta \mapsto <$

$s(\zeta), \zeta - z_0 >$ does not vanish. We now apply Theorem 2.7 with all the q_j equal to zero and s replaced by

$$\tilde{s}(\zeta) = \chi(\zeta)s(\zeta) + (1 - \chi(\zeta))\frac{\bar{f}(\zeta)}{\|f\|^2}.$$

This concludes the proof of the formula (2.16).   □

Example. If U is a strictly convex domain with a C^2 boundary and ρ a defining function for U, and $\omega \subset\subset U$, there is a strictly positive constant δ_ω such that

$$\rho(z) - \rho(\zeta) + \mathrm{Re}\,(\sum_{j=1}^{n} \frac{\partial\rho}{\partial\zeta_j}(\zeta)(\zeta_j - z_j)) \geq \delta_\omega\|\zeta - z\|^2$$

for any z in ω and any ζ in U. In such a situation, we can use (2.16) with

$$s_j(z, \zeta) = \frac{\partial\rho}{\partial\zeta_j}(\zeta)$$

and obtain a representation formula, valid for holomorphic functions in U, and such that the kernel K involved in it is holomorphic with respect to the z variables. Such is not the case for the standard choice of s, that is, when one takes $s_j(z, \zeta) = \bar{\zeta}_j - \bar{z}_j, 1 \leq j \leq n$, in which case we obtain the classical Bochner-Martinelli formula, valid in any compact domain in $\mathbb{C}^n$ with piecewise smooth boundary: for any function h holomorphic in U and continuous up to the boundary and any point z_0 inside U,

$$h(z_0) = \frac{(n-1)!(-1)^{\frac{n(n-1)}{2}}}{(2\pi i)^n} \int_{\partial U} h(\zeta) \frac{\sum_{k=1}^{k=n}(-1)^{k-1}(\bar{\zeta}_k - \bar{z}_k)d(\bar{\zeta} - \bar{z})_{[k]} \wedge d\zeta}{\|\zeta - z\|^{2n}}.$$

(2.17)

Formula (2.17) can easily be obtained from the fact that $\|\zeta\|^{2-2n}$ is, up to a constant, a fundamental solution for Laplace operator Δ in $\mathbb{R}^{2n}$.

We have the following corollary of Theorem 2.9.

Corollary 2.10. *Let U be a bounded domain in $\mathbb{C}^n$ with piecewise smooth boundary. Let $f_1, \ldots, f_n$ be n functions holomorphic in a neighborhood $\tilde{U}$ of U and s be a $\mathbb{C}^n$-valued function, C^1 on a neighborhood of ∂U, such that $s_1(\zeta)f_1(\zeta) + \ldots + s_n(\zeta)f_n(\zeta) \neq 0$ for any $\zeta \in \partial U$. Then, the set of common zeros of $f_1, \ldots, f_n$ in U is discrete and we have, for any $h \in H(U) \cap C^0(\bar{U})$ and any $z \in U$,*

$$\sum_{\substack{f_1(\alpha) = \ldots = f_n(\alpha) = 0 \\ \alpha \in U}} h(\alpha) = \frac{(n-1)!(-1)^{\frac{n(n-1)}{2}}}{(2\pi i)^n} \int_{\partial U} h(\zeta) \frac{\sum_{k=1}^{n}(-1)^{k-1}s_k(\zeta)ds_{[k]} \wedge df}{< s(\zeta), f(\zeta) >^n}.$$

(2.18)

Proof. First it is clear that $f_1, \ldots, f_n$ define a discrete variety in U, since any compact analytic variety is discrete ([13], p. 106). This implies that the set of common zeroes of $f_1, \ldots, f_n$ in U is finite. From the fact that $f_1, \ldots, f_n$ define a discrete variety, it follows that the Jacobian J of $f_1, \ldots, f_n$ cannot vanish identically in U (see for example theorem 10, [13], p.160). Because of Sard's theorem [22], the set of critical values of $\zeta \mapsto (|f_1|^2, \ldots, |f_n|^2)(\zeta)$ is a set with measure zero in $\mathbb{R}^n$. Therefore, for almost all $\epsilon_1, \ldots, \epsilon_n$, $f_1 - \epsilon_1, \ldots, f_n - \epsilon_n$ define a submanifold in U. If we prove that (2.18) holds in the case when $f_1, \ldots, f_n$ define a submanifold in U, that is, $J(f_1, \ldots, f_n)(\alpha) \neq 0$ whenever $f_1(\alpha) = \ldots = f_n(\alpha) = 0$, then, the proposition will follow in the general case through a perturbation argument. We will prove (2.18) under such an hypothesis on $f_1, \ldots, f_n$.

As in the proof of Theorem 2.9, we introduce a $\mathbb{C}^n$ valued function $\tilde{s} \in \mathcal{C}^1(\bar{U})$, which coincides with s on ∂U and satifies $< \tilde{s}(\zeta), \zeta > \neq 0$, $\zeta \in U$. An easy computation shows that the differential form

$$\frac{\sum_{k=1}^{n} (-1)^{k-1} \tilde{s}_k(\zeta) d\tilde{s}_{[k]} \wedge df_1 \wedge \ldots \wedge df_n}{< \tilde{s}(\zeta), f(\zeta) >^n}$$

is $\bar{\partial}$-closed in $U \setminus \{f_1 = \ldots = f_n = 0\}$. The implicit mapping theorem implies that the application $(f_1, \ldots, f_n)$ maps biholomorphically a ball $B(\alpha, \rho(\alpha))$ about some common zero of $f_1, \ldots, f_n$ in U onto a neighborhood of the origin in $\mathbb{C}^n$. For each such common zero α in U, there exists such a ball and such a diffeomorphism θ_α. Assume that the balls are sufficiently small so that they are pairwise disjoint, then, by Stokes's theorem

$$\int_{\partial U} h(\zeta) \frac{\sum_{k=1}^{n} (-1)^{k-1} \tilde{s}_k(\zeta) d\tilde{s}_{[k]} \wedge df}{< \tilde{s}, f >^n} =$$

$$= \sum_{\substack{f_1(\alpha)=\ldots=f_n(\alpha)=0 \\ \alpha \in U}} \int_{\partial D(\alpha, \rho(\alpha))} h(\zeta) \frac{\sum_{k=1}^{n} (-1)^{k-1} \tilde{s}_k(\zeta) d\tilde{s}_{[k]} \wedge df}{< \tilde{s}, f >^n}$$

$$= \sum_{\substack{f_1(\alpha)=\ldots=f_n(\alpha)=0 \\ \alpha \in U}} \int_{\theta_\alpha(\partial D(\alpha, \rho(\alpha)))} h(\theta_\alpha^{-1}(w)) \frac{\sum_{k=1}^{n} (-1)^{k-1} \sigma_k^{(\alpha)} d\sigma_{[k]} \wedge dw}{< \sigma_k^{(\alpha)}, w >^n}$$

where $\sigma^{(\alpha)}(w) = s(\theta_\alpha^{-1}(w))$. This sum equals

$$\frac{(2\pi i)^n (-1)^{\frac{n(n-1)}{2}}}{(n-1)!} \sum_{\substack{f_1(\alpha)=\ldots=f_n(\alpha)=0 \\ \alpha \in U}} h(\theta_\alpha(0))$$

as it can be seen from Theorem 2.9 when $z_0 = 0$. □

Corollary 2.11. *Let U be a connected open bounded set in $\mathbb{C}^n$ with piecewise C^1 boundary. Let $f_1, \ldots, f_n$ be n functions holomorphic in some neighborhood $\tilde{U}$ of U, without any common zeros on ∂U. Then, the quantity*

$$N(f, U) = \frac{(-1)^{\frac{n(n-1)}{2}}(n-1)!}{(2\pi i)^n} \int_{\partial U} \Omega(f, \varsigma) \wedge df_1 \wedge \ldots \wedge df_n,$$

where Ω denotes the differential form

$$\Omega(f, \varsigma) = \frac{\sum\limits_{k=1}^{n} (-1)^{k-1} \bar{f}_k \overline{df_{[k]}}}{\|f(\varsigma)\|^{2n}}$$

is a positive integer. Furthermore, if $f_1 = \ldots = f_n = 0$ implies $df_1 \wedge \ldots \wedge df_n \neq 0$, $N(f, U)$ is exactly the number of common zeroes of $f_1, \ldots, f_n$ in the domain U.

Proof. It is immediate to deduce this assertion from Corollary 2.9; the only thing to do is to choose $s = \bar{f}$ and to apply formula (2.18) to the function $h = 1$. The result follows. $\qquad\square$

In order to interpret this integer as the number of common zeros, counted with multiplicities, let us make a few remarks about this last corollary. We suppose $\tilde{U}$ bounded. Let $\mathcal{B}(\tilde{U})$ be the Banach algebra of functions holomorphic in $\tilde{U}$, continuous in $\bar{\tilde{U}}$, equipped with the uniform norm. We will assume that this algebra contains the f_j. Let $\eta > 0$; if ϵ is small enough, then for $g \in (\mathcal{B}(\tilde{U}))^n$, we have

$$\|f - g\| < \epsilon \Rightarrow |N(f, U) - N(g, U)| < \eta.$$

From Sard's theorem, the set of critical values of $(|f_1|^2, \ldots, |f_n|^2)$ has measure zero. Thus, we can find a sequence $(\epsilon_1^{(j)}, \ldots, \epsilon_n^{(j)})_{j \in \mathbb{N}}$ converging to zero outside this set. We have

$$N(f, U) = \lim_{j \to \infty} N(f - \epsilon^{(j)}, U).$$

From Corollary 2.10, it follows that $N(f - \epsilon^{(j)}, U)$ is precisely the number of common zeroes of $(f_i - \epsilon_i^{(j)})_{1 \leq i \leq n}$ inside U. Because N is continuous with respect to the standard norm in $\mathcal{B}(\tilde{U})$, the integer $N(f, U)$ represents the number of common distinct zeros of any perturbed system g of f, provided that all the zeros of g in $\bar{U}$ are simple. We will come back to this notion of multiplicity later on.

Theorem 2.12. (Rouché) *Let U be a connected bounded set in $\mathbb{C}^n$, with a connected piecewise C^1 boundary, and $f_1, \ldots, f_n$ be n holomorphic functions in a neighborhood of $\bar{U}$. Suppose $g_1, \ldots, g_n$ are also holomorphic in a neighborhood of $\bar{U}$ and satisfy*

$$\|g - f\| < \|f\| \text{ on } \partial U.$$

We have $N(f, U) = N(g, U)$.

Proof. One can see immediately that the function of t

$$t \mapsto \int_{\partial U} \Omega(tf + (1-t)g, \varsigma) \wedge df_1 \wedge \ldots \wedge df_n,$$

is continuous on $[0,1]$. As it takes values in $\mathbb{N}$, the value at 0 coincides with the value at 1, which means that $N(f, U) = N(g, U)$. To justify our claim, we need to remark that

$$\|tf + (1-t)g\| = \|f + (1-t)(g-f)\| \geq \|f\| - \|g-f\| \text{ on } \partial U$$

and to take into account that the differential form

$$(t, \varsigma) \mapsto \Omega(tf + (1-t)g, \varsigma)$$

remains non singular on $[0,1] \times \partial U$.   □

Theorem 2.13. (The minimum principle.) *Let U be a connected bounded set in $\mathbb{C}^n$, and $f_1, \ldots, f_n$ be n holomorphic functions in a neighborhood of $\bar{U}$. Suppose $f_1, \ldots, f_n$ have no common zeroes in $\bar{U}$, then*

$$\min_{\bar{U}} \left(\sum_{j=1}^{n} |f_j(\varsigma)|^2 \right)^{\frac{1}{2}} = \min_{\partial U} \left(\sum_{j=1}^{n} |f_j(\varsigma)|^2 \right)^{\frac{1}{2}}.$$

Remark 2.14. When $n = 1$, this minimum principle is nothing but the maximum principle applied to the function $1/f$.

Remark 2.15. The result appears as a particular case of the minimum principle for plurisubharmonic solutions of the Monge-Ampère equation. The fact that the number of functions coincides with the dimension is here crucial. For further references and a first approach of the complex Monge-Ampère equation, we refer to the work of Bedford and Taylor [4].

Proof. Let $\tilde{U}$ be a neighborhood of $\bar{U}$ such that the f_j are holomorphic and continuous up to the boundary $\partial \tilde{U}$, and consider the corresponding Banach algebra $\mathcal{B}(\tilde{U})$. Due to the hypothesis $N(f, U) = 0$, it is clear from the formula defining $N(g, U)$ that the application $g \mapsto N(g, U)$ is continuous in the open ball of $(\mathcal{B}(\tilde{U}))^n$ centered at $f = (f_1, \ldots, f_n)$ and with radius $\rho = \min\{\|f(\varsigma)\|, \ \varsigma \in \partial U\}$. Then, as $g \mapsto N(g, U)$ takes its values in $\mathbb{N}$, $N(g, U) = 0$. In particular, for any $\eta \in \mathbb{C}^n$ such that $\|\eta\| < \rho$, $N(f - \eta, U) = 0$. This shows that $(f_1, \ldots, f_n)$ cannot take in U the value $(\eta_1, \ldots, \eta_n)$ whenever $\|\eta\| < \rho$. It follows that

$$\min_{\partial U} \|f\| = \min_{\bar{U}} \|f\|.$$

This concludes the proof of the minimum principle.   □

§4 Weighted Andreotti-Norguet formulas

A variant of Cauchy's formula for test functions can be obtained by the procedure developed in the preceding sections. For any n-uplet $(\beta_1, \ldots, \beta_n)$ in $\mathbb{N}^n$, and any $\phi \in \mathcal{D}(\mathbb{C}^n)$, we have

$$\frac{\partial^{|\beta|}}{\partial z^\beta} \phi(z) = \beta! \lim_{\lambda \to \underline{0}} \frac{(-1)^{\frac{n(n-1)}{2}}}{(2\pi i)^n} \int |\zeta - z|^{*2(\beta + \underline{1})(\lambda - \underline{1})} \overline{\partial(\zeta - z)^{\beta + \underline{1}}} \wedge \phi d\zeta. \quad (2.19)$$

in which the limit above means that one takes the analytic continuation and then the value at $\lambda = \underline{0}$) (see proposition 2.2). Formula (2.19) allows us to redo the proof of Theorem 2.7, using $(\zeta_1 - z_1)^{\beta_1 + 1}, \ldots, (\zeta_n - z_n)^{\beta_n + 1}$ instead of $(\zeta_1 - z_1), \ldots, (\zeta_n - z_n)$. We consider as in the preceding section a bounded domain U with piecewise smooth boundary, an open set ω which is such $\omega \subset\subset U$. Here $(\beta_1, \ldots, \beta_n)$ will be kept fixed. We consider a collection of M $\mathbb{C}^n$-valued functions $q_1, \ldots, q_M$ which are in $\mathcal{C}^1(\bar\omega \times \bar U)$. We also introduce M functions of one variable $G_1, \ldots, G_M$, such that G_k is defined and holomorphic near the image of $\bar\omega \times \bar U$ by the mapping

$$(z, \zeta) \mapsto 1 - \sum_{j=1}^{n} q_{k,j}(z, \zeta)(\zeta_j - z_j)^{\beta_j + 1} = 1 - < q_k, (\zeta - z)^{\beta + \underline{1}} >$$

and satisfies $G_k(1) = 1$. Remember that, despite a slight ambiguity in the notations, for any ζ in $\mathbb{C}^n$ and any $r \in \mathbb{N}^n$, the element ζ^r will denotes the element in $\mathbb{C}^n$ whose coordinates are the $\zeta_k^{r_k}$, $1 \le k \le n$ (it has to be differentiated from $\zeta^{*r} = \zeta_1^{r_1} \cdots \zeta_n^{r_n}$. For the sake of simplicity, we denote

$$\Gamma_{k,\beta}^{(\alpha)}(z, \zeta) = \frac{d^\alpha}{dt^\alpha} G_k|_{t = 1 - <q_k, (\zeta - z)^{\beta + \underline{1}}>}, \ 1 \le k \le M, \quad \text{and } \alpha \in \mathbb{N}.$$

As before, s is a $\mathbb{C}^n$-valued mapping from $\bar\omega \times \bar U$ into $\mathbb{C}^n$, of class $\mathcal{C}^1$, which is such that the function

$$(z, \zeta) \mapsto < s(z, \zeta), (\zeta - z)^{\beta + \underline{1}} >$$

does not vanish on $\omega \times \partial U$. The notations $Q_0, \bar\partial Q_0,, S, \bar\partial S$, are analogous to those in §3. We have the following representation formulas for derivatives of holomorphic functions

Theorem 2.16. *Let h be holomorphic in U, $\mathcal{C}^1$ in $\bar U$. Let ω, $\beta \in \mathbb{N}^n$, $q_1, \ldots, q_M$, $G_1, \ldots, G_M$, s as above. Then for $z \in \omega$ we have*

$$\frac{\partial^{|\beta|}}{\partial z^\beta} h(z) = \frac{\beta!}{(2\pi i)^n} \left(\int_U h(\zeta) P_\beta(z, \zeta) + \int_{\partial U} h(\zeta) K_\beta(z, \zeta) \right). \quad (2.20)$$

with

$$P_\beta(z,\zeta) = \sum_{|\alpha|=n} \frac{1}{\alpha!} \Gamma_{1,\beta}^{(\alpha_1)} \dots \Gamma_{M,\beta}^{(\alpha_M)} \bar{\partial} Q_1^{\alpha_1} \wedge \dots \wedge \bar{\partial} Q_M^{\alpha_M}(z,\zeta)$$

$$K_\beta(z,\zeta) = \sum_{\alpha_0+|\alpha|=n-1} \frac{1}{\alpha!} \Gamma_{1,\beta}^{(\alpha_1)} \dots \Gamma_{M,\beta}^{(\alpha_M)} \frac{S \wedge (\bar{\partial} S)^{\alpha_0} \wedge \bar{\partial} Q_1^{\alpha_1} \wedge \dots \wedge \bar{\partial} Q_M^{\alpha_M}}{< s(z,\zeta), (z-\zeta)^{\beta+\underline{1}} >^{\alpha_0+1}}(z,\zeta)$$

The classical Norguet-Andreotti [19] formula is obtained from Theorem 2.16 by letting $q_1 = \dots = q_M = 0$, its statement is as follows.

Theorem 2.17. *Let U be a bounded domain in $\mathbb{C}^n$ with piecewise C^1 boundary. Let $\beta \in \mathbb{N}^n$ and s be a $\mathbb{C}^n$-valued function of class C^1 in a neighborhood of ∂U such that $< s(z,\zeta), (\zeta-z)^{\beta+\underline{1}} > \neq 0$ for any $\zeta \in \partial U$ and some $z_0 \in U$. In this situation, any $h \in H(U) \cap C(\bar{U})$ one has the following representation*

$$\frac{\partial^{|\beta|}}{\partial z^\beta} h(z_0) = \frac{\beta!(n-1)!(-1)^{\frac{n(n-1)}{2}}}{(2\pi i)^n} \int_{\partial U} h(\zeta) \frac{\sum_{k=1}^{n} (-1)^{k-1} s_k(\zeta) ds_{[k]} \wedge d\zeta}{< s(\zeta), (\zeta-z_0)^{\beta+\underline{1}} >^n}. \quad (2.21)$$

As in Section 3, we have the following corollary.

Corollary 2.18. *Let U be a bounded domain in $\mathbb{C}^n$ with piecewise C^1 boundary. Let $f_1, \dots, f_n$ be n functions holomorphic in a neighborhood of $\bar{U}$, whose only common zeros in $\bar{U}$ lie inside U and are all simple. Let s be a $\mathbb{C}^n$-valued function of class C^1 in a neighborhood of ∂U such that $s_1 f_1^{\beta_1+1} + \dots + s_n f_n^{\beta_n+1} \neq 0$ on the boundary of U. Hence, for any $h \in H(U) \cap C(\bar{U})$ we have*

$$\sum_{\substack{\alpha \in U \\ f_1(\alpha)=\dots=f_n(\alpha)=0}} \frac{\partial^{|\beta|}}{\partial z^\beta} h(\alpha) = C_1(\beta, n) \int_{\partial U} h(\zeta) \frac{\sum_{k=1}^{n} (-1)^{k-1} s_k(\zeta) ds_{[k]} \wedge df}{< s(\zeta), f(\zeta)^{\beta+\underline{1}} >^n}, \quad (2.22)$$

where the constant $C_1(\beta, n)$ is given by

$$C_1(\beta, n) = \frac{\beta!(n-1)!(-1)^{\frac{n(n-1)}{2}}}{(2\pi i)^n}.$$

The proof of this corollary is exactly the same as that of Corollary 2.10 . We can weaken the hypothesis that the zeros of $f_1, \dots, f_n$ are simple if we use the concept of ordinary point due to Bolotov (see Proposition 5.4 in [1], p.42.) In any case the main problem in (2.18) or (2.22) remains the fact that df is in the integrand. Our goal in Chapter 3 will be to study what happens when df is replaced by $d\zeta$.

Remark 2.19. It is important to note that there is another way to express $\frac{\partial^{|\beta|}}{\partial z^\beta}h(z_0)$ than in the Theorem 2.16. One can use Theorem 2.9 and just differentiate under the integral, so that for the choice $s(\zeta, z) = \bar\zeta - \bar z$ one obtains

$$\frac{\partial^{|\beta|}}{\partial z^\beta}h(z_0) = C_2(\beta, n) \int_{\partial U} h(\zeta) \frac{\left(\sum_{k=1}^{n}(-1)^{k-1}(\bar\zeta_k - \bar z_{0,k})(\zeta - z_0)^{*\beta}d\bar\zeta_{[k]}\right) \wedge d\zeta}{\|\zeta - z_0\|^{2(|\beta|+n)}}.$$

$$(2.23)$$

with another constant $C_2(\beta, n)$ which is given by

$$C_2(\beta, n) = \frac{(|\beta| + n - 1)!(-1)^{\frac{n(n-1)}{2}}}{(2\pi i)^n}.$$

We now deduce from (2.23), assuming the same hypothesis than the one we had in Corollary 2.18, introduce a complex parameter μ and get

$$\sum_{\substack{\alpha \in U \\ f_1(\alpha)=...=f_n(\alpha)=0}} \frac{\partial^{|\beta|}}{\partial z^\beta}h(\alpha) =$$

$$= C_2(\beta, n) \lim_{\substack{\mu \to 0 \\ \text{Re}\,\mu > 0}} \left(\int_{\partial U} h(\zeta)|f_1 \ldots f_n|^{2(n+|\beta|)\mu} \bar f^{*\beta} \frac{\Omega(f, \zeta)}{\|f\|^{2|\beta|}} \wedge df \right)$$

$$= C_2(\beta, n) \lim_{\substack{\mu \to 0 \\ \text{Re}\,\mu > 0}} \left(\int_{U} h(\zeta) \left(\bar\partial[|f_1 \ldots f_n|^{2(n+|\beta|)\mu}] \right) \wedge \bar f^{*\beta} \frac{\Omega(f, \zeta)}{\|f\|^{2|\beta|}} \wedge df \right)$$

where we recall

$$\Omega(f, \zeta) = \frac{\sum_{k=1}^{n}(-1)^{k-1}\bar f_k \overline{df_{[k]}}}{\|f\|^{2n}}$$

and the limit at $\mu = 0$ is understood as the value at $\mu = 0$ of the analytic continuation. On the other hand, from (2.22) with $s(\zeta) = f(\zeta)^{\beta+\underline{1}}$, we conclude that

$$\sum_{\substack{\alpha \in U \\ f_1(\alpha)=...=f_n(\alpha)=0}} \frac{\partial^{|\beta|}}{\partial z^\beta}h(\alpha) =$$

$$= C_1(\beta, n) \lim_{\substack{\mu \to 0 \\ \text{Re}\,\mu > 0}} \left(\int_{\partial U} h(\zeta)|f(\zeta)|^{*2\mu(\beta+\underline{1})} \Omega_\beta(f, \zeta) \wedge df \right)$$

$$= C_1(\beta, n) \lim_{\substack{\mu \to 0 \\ \text{Re}\,\mu > 0}} \left(\int_{U} h(\zeta) \left(\bar\partial[|f|^{*2\mu(\beta+\underline{1})}] \right) \wedge \Omega_\beta(f, \zeta) \wedge df \right)$$

where the differential form Ω_β is defined as

$$\Omega_\beta(f, \zeta) = \frac{\sum_{k=1}^n (-1)^{k-1} \bar{f}_k^{\beta_k+1} \overline{df_{[k]}^{\beta+\underline{1}}}}{\|f^{\beta+\underline{1}}\|^{2n}}$$

We summarize these computations in the following statement.

Proposition 2.20. *Let U be a bounded domain in $\mathbb{C}^n$ with a piecewise $\mathcal{C}^1$ boundary. Let $f_1, \ldots, f_n$ be n functions holomorphic in a neighborhood of $\bar{U}$ without any common zero on ∂U. Let $\beta \in \mathbb{N}^n$ and $h \in H(U) \cap \mathcal{C}(\bar{U})$. The two functions of one complex variable $\mu \mapsto J_1(\mu, f, h)$, and $\mu \mapsto J_2(\mu, f, h)$ defined for $\mathrm{Re}\ \mu \gg 0$ by*

$$J_1(\mu, f, h) = \mu(n + |\beta|)! \int_U h(\zeta) |f_1 \ldots f_n|^{2(n+|\beta|)\mu} \frac{\bar{f}^\beta}{\|f\|^{2(n+|\beta|)}} d\bar{f} \wedge df$$

$$J_2(\mu, f, h) = \mu n! \beta! \int_U h(\zeta) |f|^{*2\mu(\beta+\underline{1})} \frac{\overline{df^{\beta+\underline{1}}} \wedge df}{\|f^{(\beta+\underline{1})}\|^{2n}}$$

can be analytically continued as functions of μ to be holomorphic in $\mathrm{Re}\ \mu > 0$, continuous in $\mathrm{Re}\ \mu \geq 0$. Moreover, $J_1(0, f, h) = J_2(0, f, h)$.

We will later on in Chapter 3 interpret the common value of $J_1(0, f, h) = J_2(0, f, h)$ in terms of the action on h of the residue current attached to the holomorphic functions $f_1^{\beta_1+1}, \ldots, f_n^{\beta_n+1}$. We will also be able there to replace df by $d\zeta$ and even to prove the same result when h is continuous in $\bar{U}$ and $\mathcal{C}^\infty$ near the common zeroes of $f_1, \ldots, f_n$ in U.

§5 Applications to systems of algebraic equations

The ideas we develop in this section are inspired from the work of L.A. Aizenberg and A. Tsikh [1], [2]. The reason why we introduce them here is to motivate further our approach to effectivity questions in commutative algebra.

Consider n polynomials in n variables of the form :

$$P_k(z) = z_k^{m_k} + Q_k(z), \quad 1 \leq k \leq n,$$

where the m_k are strictly positive integers and the Q_k some polynomials in n variables which satisfy

$$\deg(Q_1) < m_1, \ldots, \deg(Q_n) < m_n.$$

Such polynomials define a compact, hence a discrete variety V in $\mathbb{C}^n$. Moreover there are no common zeroes at infinity (the only common zero of the homogeneous parts of higher degree of the P_k is the origin in $\mathbb{C}^n$.) We introduce for $r > 0$, the pseudoball

$$U_r^m = \{\zeta \in \mathbb{C}^n,\ |\zeta_1|^{2m_1} + \ldots + |\zeta_n|^{2m_n} = r^2\}.$$

For $r \gg 0$, this pseudoball contains all common zeroes of $P_1, \ldots, P_n$ (the set $\{P_1 = \ldots = P_n = 0\}$ is finite, such is the case for any 0-dimensional algebraic variety in $\mathbb{C}^n$ as we will see later on.) Before stating our first result, we need a few notations.

Notations. We will denote as $\mathcal{N}$ the linear functional defined over the vector space of polynomials in $\zeta_1, \ldots, \zeta_n, \bar{\zeta}_1^{m_1}, \ldots, \bar{\zeta}_n^{m_n}$ by the following rules: if α, β are two elements in $\mathbb{N}^n$,

$$\mathcal{N}(\zeta^\beta \bar{\zeta}^{m\alpha}) = \begin{cases} \alpha_1! \ldots \alpha_n!, & \text{if } \beta_j = m_j \alpha_j + m_j - 1, \ j = 1, \ldots n \\ 0 & \text{otherwise.} \end{cases}$$

The polynomial J wil be the Jacobian of the polynomial map $(P_1, \ldots, P_n)$ and the $\mathbb{C}^n$-valued application s will be defined by

$$s(\zeta) = (\bar{\zeta}_1^{m_1}, \ldots, \bar{\zeta}_n^{m_n}).$$

Therefore, for any positive integer j, for any polynomial R in $\zeta_1, \ldots, \zeta_n$, the expression $R \times J \times < s, Q >^j$ will be considered as a polynomial in $\zeta_1, \ldots, \zeta_n$, $\bar{\zeta}_1^{m_1}, \ldots, \bar{\zeta}_n^{m_n}$. The action of $\mathcal{N}$ on it will make sense. We now can state the following proposition.

Proposition 2.21. *Let $P_1, \ldots, P_n$ be as above and R be a polynomial in n variables of total degree at most $k \in \mathbb{N}$. Then*

$$\sum_{\alpha \in V} R(\alpha) = \mathcal{N}\left[RJ \sum_{p=0}^{k} \frac{(-1)^p}{p!} < s, Q >^p \right]. \tag{2.24}$$

Proof. We use formula (2.18) with $U = U_r^m$ and $s(\zeta) = (\bar{\zeta}_1^{m_1}, \ldots, \bar{\zeta}_n^{m_n})$ as defined above. For $\zeta \in \partial U_r^m$, one can use the following expansion of

$$\frac{1}{< s(\zeta), P(\zeta) >^n} = \frac{1}{\left(r^2 \left(1 + \frac{<s(\zeta), Q(\zeta)>}{r^2} \right) \right)^n},$$

namely,

$$\frac{1}{< s(\zeta), P(\zeta) >^n} = \sum_{p=0}^{\infty} (-1)^p \frac{(p + n - 1)!}{p!} \frac{< s(\zeta), Q(\zeta) >^p}{r^{2(p+n)}}. \tag{2.25}$$

Let, for $r > 0$, $\alpha \in \mathbb{N}^n$, $\beta \in \mathbb{N}^n$

$$I_{r,\alpha,\beta} = (-1)^{\frac{n(n-1)}{2}} \frac{1}{(2\pi i)^n} \int_{\partial U_r^m} \zeta^{*\beta} \bar{\zeta}^{*m\alpha} \sum_{j=1}^{n} (-1)^{j-1} \bar{\zeta}_j^{m_j} d\bar{\zeta}_{[j]}^m \wedge d\zeta.$$

Such an integral can be transformed with the help of Stokes's theorem into

$$I_{r,\alpha,\beta} = (-1)^{\frac{n(n-1)}{2}} \frac{n+|\alpha|}{(2\pi i)^n} \int_{U_r^m} \zeta^{*\beta} \bar\zeta^{*m\alpha} d\bar\zeta^m \wedge d\zeta.$$

Because of the invariance of U_r^m under the change of variables

$$\zeta_1 = e^{it_1}\zeta_1', \dots, \zeta_n = e^{it_n}\zeta_n'$$

for any $(t_1, \dots, t_n) \in [0, 2\pi]^n$, it is clear that $I_{r,\alpha,\beta} = 0$ unless

$$\beta_j = m_j(\alpha_j + 1) - 1, \quad j = 1, \dots, n.$$

When such conditions are fulfilled, one has

$$I_{r,\alpha,\beta} = (|\alpha| + n) \left(\int_{\tau_1 + \dots + \tau_n \leq 1} \tau_1^{\alpha_1} \dots \tau_n^{\alpha_n} d\tau_1 \dots d\tau_n \right) r^{2(|\alpha|+n)}$$

$$= \frac{\alpha!}{(|\alpha| + n)!} r^{2(|\alpha|+n)}.$$

Let now Ω_s be the differential form

$$\Omega_s(\zeta) = \sum_{j=1}^n (-1)^{j-1} \bar\zeta_j^{m_j} d\bar\zeta_{[j]}^m \wedge dP.$$

Formula (2.18) gives us, for large r,

$$\sum_{\alpha \subset V} R(\alpha) = \frac{(n-1)!(-1)^{\frac{n(n-1)}{2}}}{(2\pi i)^n} \int_{\partial U_r^m} R(\zeta) \frac{\Omega_s(\zeta)}{< s(\zeta), P(\zeta) >^n},$$

which we transform first by developing the integral kernel by means of formula (2.25), then by transforming the $I_{r,\alpha,\beta}$ using Stokes's theorem. Formula (2.24) is then a consequence of the computations we perfomed above to compute the $I_{r,\alpha,\beta}$. The only last point to remark is the fact that, for $\alpha \in \mathbb{N}$, the total degree in $\zeta_1, \dots, \zeta_n$ of the expression $RJ(\bar\zeta_1^{m_1} Q_1)^{\alpha_1} \dots (\bar\zeta_n^{m_n} Q_1)^{\alpha_n}$ is at most $k + |m| - n + m_1\alpha_1 + \dots + m_n\alpha_n - |\alpha|$. This implies the action of the functional $\mathcal{N}$ on such a term is zero as soon as $|\alpha| > k = \deg(R)$. This shows that formula (2.18) becomes here (2.24) and concludes the proof of the proposition. □

Note that Proposition 2.21 implies Bezout theorem in the affine setting:

Theorem 2.22. *If $P_1, \dots, P_n$ define a discrete variety in $\mathbb{C}^n$, then the number of common zeroes of $P_1, \dots, P_n$, counted with their multiplicities, is at most $\deg(P_1) \cdots \deg(P_n)$.*

Proof. Let e be an integer larger than 1. Given some complex parameter $\lambda \in \mathbb{C}^*$, let us introduce the polynomials

$$\zeta \mapsto P_{j,\lambda}(\zeta) = \zeta_j^{e\deg(P_j)+1} + \frac{1}{\lambda} P_j^e, \ j = 1, \ldots, n.$$

Let $\mathcal{M}$ be the linear functional defined on the vector space of Laurent polynomials in ζ which associates to any Laurent polynomial in n variables its free term. Let J_λ be the Jacobian of the polynomial map P_λ. Applying (2.24) with $R = 1$ and the $P_{j,\lambda}$, we see that

$$\text{card}\{\zeta, P_{j,\lambda}(\zeta) = 0, j = 1, \ldots, n\} = \mathcal{M}\left[\frac{\zeta_1 \cdots \zeta_n}{\zeta_1^{\deg(P_{1,\lambda})} \cdots \zeta_n^{\deg(P_{n,\lambda})}} J_\lambda(\zeta)\right]$$

$$= \deg(P_{1,\lambda}) \cdots \deg(P_{n,\lambda}).$$

We will denote this cardinal $N(\lambda)$. When the complex parameter λ tends to 0, one can see, as a consequence of Rouché's theorem (namely Theorem 2.12) the following fact: if N denotes the number of common zeros of the P_j, then $N(\lambda) - N$ common zeros of the $P_{j,\lambda}$ tend to infinity, while the others approach the common zeros of $P_1, \ldots, P_n$. Therefore we have

$$\text{card}\{\zeta, P_j(\zeta) = 0, j = 1, \ldots, n\} \leq (e\deg(P_1) + 1) \ldots (e\deg(P_n) + 1). \quad (2.26)$$

If we divide (2.26) by e^n and if we take e large enough, then we get the desired conclusion, that is Bezout theorem. $\square$

As a corollary of this result, let us make the following remark: if $P_1, \ldots, P_n$ define locally the origin as an isolated zero, then , for a generic choice of ϵ, $P_1 - \epsilon_1, \ldots, P_n - \epsilon_n$ have only non critical common zeroes (this is just a consequence of Sard's theorem) and therefore define a discrete variety in $\mathbb{C}^n$; Theorem 2.22 applies to $P_1 - \epsilon_1, \ldots, P_n - \epsilon_n$ and we have

$$\text{mult}_0\{P_1 = \ldots = P_n = 0\} \leq \deg(P_1) \cdots \deg(P_n). \quad (2.27)$$

Here the multiplicity at the origin is the integer we have introduced in Corollary 2.11.

Remark 2.23. Suppose that $P_1, \ldots, P_n$ are n elements in $\mathbb{C}[X_1, \ldots, X_n]$ defining a discrete algebraic variety in $\mathbb{C}^n$. The cardinal of the zero set of $(P_1, \ldots, P_n)$ is at most $\deg(P_1) \cdots (\deg P_n)$; the main problem is how to compute these zeros. Let us recall here two methods to do that. The first one is to use elimination theory. As $P_1, \ldots, P_n$, considered as polynomials in $\mathbb{C}[X_1][X_2, \ldots, X_n]$, define an empty variety in $\mathbb{K}^n$, where $\mathbb{K}$ denotes some algebraic closure of $\mathbb{C}[X_1]$, one can find some polynomial $Q_1 \in \mathbb{C}[X_1]$ and $A_1, \ldots, A_n$ in $\mathbb{C}[X_1, \ldots, X_n]$ such that, for any ζ in $\mathbb{C}^n$,

$$Q_1(\zeta_1) = A_1(\zeta)P(\zeta) + \ldots + A_n(\zeta)P(\zeta). \quad (2.28)$$

One can also construct $Q_2(X_2), \ldots, Q_n(X_n)$ satisfying analogous properties with respect to the other variables $X_2, \ldots X_n$. The problem of this method is that, though generally it works effectively by using Groebner bases as in [8], one does not effective bounds for its complexity, because finding estimates involves computations of resultants (compare to the algorithm of G. Hermann [14], rewritten for example in [18], section 4). The bound on the degree one can expect for $Q_1, \ldots, Q_n$ using such an algorithm is $K(n)D^{O(2^n)}$, if D is the maximum of the degrees of the polynomials P_j. We will see in chapter 4 that one can prove the existence of polynomials Q_j which can be written as (2.28) and are such that $\deg(Q_1), \ldots, \deg(Q_n)$ have bounds of the form

$$\deg(Q_j) \leq K(n)D^{O(n)}.$$

These bounds are, of course, much better than the doubly exponential (see [16], [20], [21]); it remains the delicate problem to get algorithms in order to realize such economic division formulas; some recent results (which we will quote in chapter 4), following the pioneer work of Grigoriev and Chistov ([9], [12]), constitute some encouraging attempt in order to solve, in an algorithmic way and in subexponential time, the problem of finding equations for the irreducible components of an algebraic variety.

The second idea we want to develop here is totally different. Suppose $P_1, \ldots, P_n$ define a discrete variety. Let $D_1 = \deg(P_1), \ldots, D_n = \deg(P_n)$. Let us come back to the proof of Theorem 2.22 where we introduced our complex parameter λ. Let $\lambda \in \mathbb{C}^*$ and

$$\tilde{P}_{j,\lambda}(\zeta) := \zeta_j^{D_j+1} + \frac{1}{\lambda}P_j(\zeta).$$

For any value of $\lambda \in \mathbb{C}^*$, $\tilde{P}_{1,\lambda}, \ldots, \tilde{P}_{n,\lambda}$ have exactly $(D_1+1)\cdots(D_n+1) = M$ common zeroes (counted with multiplicities). We can always assume that $\underline{0}$ is not a common zero of $P_1, \ldots, P_n$. When $\lambda \to 0$, it follows as before from Rouché's Theorem that if $\text{card}(\{P_1 = \ldots P_n = 0\}) = N$, $M - N$ roots among the roots (that we will call $\zeta_j^{(\lambda)}, j = 1, \ldots, M$) of $\tilde{P}_{1,\lambda}, \ldots, \tilde{P}_{n,\lambda}$ tend to infinity, while the other ones converge to the common zeroes of $(P_1, \ldots, P_n)$. If we take some generic linear form L (here is the weak point of this method, namely how to choose L, in particular such that it does not vanish at the common zeros of the P_j and such that, if $\zeta_j^{(\lambda)} \to \infty$, then $L(\zeta_j^{(\lambda)}) \to \infty$), one can compute, with formula (2.24), the Newton sums $\Sigma^1(\lambda), \ldots, \Sigma^M(\lambda)$ of the numbers $L(\zeta_1^{(\lambda)}), \ldots, L(\zeta_M^{(\lambda)})$. Then the symmetric functions $\sigma^1(\lambda), \ldots, \sigma^M(\lambda)$ of these numbers may be computed using the standard recurrent formulas. We now consider

$$\left(\frac{\sigma^{M-1}(\lambda)}{\sigma^M(\lambda)}, \frac{\sigma^{M-2}(\lambda)}{\sigma^M(\lambda)}, \ldots, \frac{\sigma^1(\lambda)}{\sigma^M(\lambda)}, \frac{1}{\sigma^M(\lambda)}\right) = \Phi(\lambda).$$

From the hypotheses on L,

$$\lim_{\lambda \to 0} \Phi(\lambda) = (\sigma_{(L)}^1, \ldots, \sigma_{(L)}^N, 0, \ldots, 0).$$

where the $\sigma^j_{(L)}$, $j = 1, \ldots N$ are the elementary symmetric polynomials in the $L(\xi_j)$, the ξ_j being the common zeroes of $(P_1, \ldots, P_n)$. It is interesting to note here that if $P_1, \ldots, P_n$ have integer coefficients, and if one chooses L with integer coefficients, then the $\sigma^j_{(L)}$ are rational numbers whose numerators and denominators can be estimated in terms of the degrees of $P_1, \ldots, P_n$, the maximum modulus of all coefficients involved in $P_1, \ldots, P_n$, and (unfortunately!) in the generic linear form L. We plan to discuss these questions in Chapter 4.

References for Chapter 2

[1] L.A. Aizenberg and A.P. Yuzhakov, *Integral Representation in Multidimensional Complex Analysis,* Translations of Amer. Math. Soc. 58, 1980.

[2] L.A. Aizenberg and A.K. Tsikh, Application of the multidimensional logarithmic residue to systems of nonlinear algebraic equations, Sibirsk. Math. Zh. 20(1979), 699–707; English transl. in Siberian Math. Journal, 20 (1979).

[3] M. Andersson and M. Passare, A shortcut to weighted representation formulas for holomorphic functions, Ark. Math. 26(1988), 1–12.

[4] E. Bedford and B.A. Taylor, The Dirichlet problem for a complex Monge-Ampère equation, Inventiones Math. 37(1976), 1–44.

[5] C.A. Berenstein and A. Yger, Effective Bezout identities in $\mathbb{Q}[z_1, \ldots, z_n]$, Acta Mathematica, 166(1991), 69–120.

[6] B. Berndtsson and M. Andersson, Henkin-Ramirez formulas with weight factors, Ann. Inst. Fourier 32(1982), 91–110.

[7] B. Berndtsson, Weighted integral formulas. Colloquium lecture at the Mittag-Leffler Institute. 1988.

[8] B. Buchberger, An algorithmic method in polynomial ideal thory, in *Multi Systems Theory,* (ed N. Bose), Reidel Publ, Dordrecht,1985.

[9] A.L. Chistov, An algorithm of polynomial complexity for factoring polynomials and finding the components of varieties in subexponential time, Journal of Soviet Math. 34(1986), 1838–1882.

[10] J.P. Demailly and C. Laurent-Thiébaut, Formules intégrales pour les formes différentielles de type (p,q) dans les variétés de Stein, Ann. Sci. Ec. Norm. Sup. Paris 20(1987), 579–598.

[11] G. De Rham, *Variétés différentiables, Formes, Courants, Formes harmoniques,* Actualités Sci.Indust. 1222, Hermann, Paris, 1955.

[12] D.Y. Grigoriev and A.L. Chistov, Fast decomposition of polynomials into irreducible ones and the solution of algebraic equations, Soviet. Math. Dokl. 29(1984), 380–383.

[13] R. Gunning and H. Rossi, *Analytic functions of several complex variables,* Prentice Hall, Englewood Cliffs, New Jersey,1965.

[14] G. Hermann, Die Frage der endliche vielen Schritte in der Theorie der Polynomideale, Math. Ann. 95(1926), 737–788.

[15] L. Hörmander, *An Introduction to Complex Analysis In Several Complex Variables,* North Holland, Amsterdam, 1966.

[16] J. Kollár, Sharp effective Nullstellensatz, Journal of Amer. Math. Soc. 1(1988), 963–975.

[17] J. Leiterer and G. Henkin, Global integral formulas for solving the $\bar{\partial}$ equation, Ann. Pol. Math. 39(1981), 93–116.

[18] D.W. Masser and G. Wüstholz, Fields of large transcendence degree generated by values of elliptic functions, Inventiones. Math. 72(1983), 407–464.

[19] F. Norguet, *Introduction aux fonctions de plusieurs variables complexes, représentations intégrales,* Lecture Notes in Math. 409(1974), Springer Verlag, New York, 1–97.

[20] P. Philippon, Théorème des zéros effectif, d'après Kollár, Publ. Math. de l'Univ. P.et M.Curie, Paris VI, 88, Groupe d'étude sur les Problèmes diophantiens 1987–88, exposé 6.

[21] P. Philippon, Dénominateurs dans le théorème des zéros de Hilbert, Acta Arithmetica 58(1991), 1–25.

[22] A. Sard, The measure of the critical values of differentiable maps, Bull. Amer. Math. Soc. 48(1942), 883–890.

[23] B.L. Van der Waerden, *Algebra*, Springer Verlag, New York 1969.

Chapter 3
Residue Currents and Analytic Continuation

§1 Leray iterated residues

From the computational point of view, the theory of multidimensional residues arises as follows. Given a complex n-dimensional manifold $\mathcal{X}$, equipped with an orientation, and a closed subset $\mathcal{S}$, we consider a d-closed p-form ϕ, smooth on $\mathcal{X} \setminus \mathcal{S}$, singular on $\mathcal{S}$, and we wish to compute $\int_\gamma \phi$ where γ is an element in $Z_p(\mathcal{X} \setminus \mathcal{S})$. Such an integral depends only on the homology class of γ in $H_p(\mathcal{X} \setminus \mathcal{S})$ and on the cohomology class of ϕ in $H^p(\mathcal{X} \setminus \mathcal{S})$. The two main cases studied by Leray are the case where $\mathcal{S}$ is a $n-1$ dimensional submanifold of $\mathcal{X}$ and the case where $\mathcal{S} = \mathcal{S}_1 \cup \ldots \cup \mathcal{S}_M$, where $\mathcal{S}_j$, $j = 1, \ldots, M$ are $n-1$ dimensional submanifolds of $\mathcal{X}$ in general position, this means that, for any set of distinct indices $j_1, \ldots, j_k$, where $1 \leq k \leq M$, for any point a in $\mathcal{S}_{j_1} \cap \ldots \cap \mathcal{S}_{j_k}$, there is a system of local coordinates about a, such that in these coordinates, $\mathcal{S}_{j_l} = \{s_{j_l}(\zeta) = 0\}$, $l = 1, \ldots, k$, and the s_{j_l} are holomorphic functions such that

$$\mathrm{grad}(s_{j_1})(a) \wedge \ldots \wedge \mathrm{grad}(s_{j_k})(a) \neq 0.$$

In the first case, the result of Leray [28] can be stated as follows.

Theorem 3.1. *Let $\mathcal{S}$ be a $n-1$ dimensional submanifold in a complex analytic manifold $\mathcal{X}$. There exists a coboundary operator δ, which induces for each $p \in \mathbb{N}$ an homomorphism δ_p from $H_p(\mathcal{S})$ to $H_{p+1}(\mathcal{X} \setminus \mathcal{S})$ and a residue operator Res, which induces for each $p \in \mathbb{N}$ an homomorphism Res $_p$ from $H^{p+1}(\mathcal{X} \setminus \mathcal{S})$ to $H^p(\mathcal{S})$ satisfying*

$$\forall \sigma \in Z_p(\mathcal{S}), \ \forall \phi \in Z^{p+1}(\mathcal{X} \setminus \mathcal{S}) \quad \int_\gamma \phi = 2i\pi \int_\sigma \psi, \tag{3.1}$$

where γ is any representative of $\delta_p[\sigma]$ and ψ any representative of Res $_p[\phi]$.

The details of the proof of this theorem can be found in ([1], Chapter 3). We will just mention here how δ and Res can be defined.

First δ is defined as follows: we assume from now on that $\mathcal{X}$ has been equipped with a Hermitian metric; let $\sigma \in Z_p(\mathcal{S})$; an element $\gamma \in Z_{p+1}(\mathcal{X} \setminus \mathcal{S})$ is in the homology class $\delta[\sigma]$ if and only if γ is homologous in $\mathcal{X} \setminus \mathcal{S}$ to $\partial\tau$, where τ is a $p+2$ smooth chain in $\mathcal{X}$ intersecting transversally $\mathcal{S}$ along a cycle in $Z_p(\mathcal{S})$, which is homologous in $\mathcal{S}$ to σ. Note that any orientation for a k-simplex σ in $\mathcal{S}$ induces, jointly with the given orientation of $\mathcal{X}$, an orientation for the $k+2$ simplex $\cup_{\alpha \in \sigma} V_\alpha$, where V_α is the union of geodesic segments issuing from α, orthogonal to $\mathcal{S}$, whose length $\rho(\alpha)$ remains sufficiently small (note that the function $\alpha \mapsto \rho(\alpha)$ can be assumed to be $\mathcal{C}^1$ on $\mathcal{S}$.) In the construction of δ (see for example [1], p.102), one extensively uses the existence of a retract from a neighborhood of S, of the form $V = \cup_{\alpha \in S} V_\alpha$, on S itself; the existence of such a retract follows from the fact that S is a submanifold of X (see for example [1], Chapter 1).

The residue operator could be defined using a duality argument based on De Rham's Theorem (see Theorem 2.1 above). More interesting for us is the explicit construction of Res $_p[\phi]$ when $[\phi]$ admits a representative ϕ that is *semi-meromorphic in* $X \setminus S$, *with a pole of order* k *on* S (this means that exits $k \in \mathbb{N}^*$ such that, locally, $s^k \phi$ is a C^∞ form, s being an holomorphic function defining the submanifold S.) We need first to prove a lemma on division of differential forms.

Lemma 3.2. *Suppose* U *is an open set in* $\mathbb{C}^n$, s *an element in* $H(U)$ *such that* $\mathrm{grad}(s) \neq 0$ *in* U. *Let* $p \in \mathbb{N}$; *for any semi-meromorphic form* ϕ *in* $Z^{p+1}(U \setminus \{s = 0\})$, *with a pole of order* $k \geq 1$ *on* $\{s = 0\}$, *there are* C^∞ *forms* ψ *and* θ *in* U *such that*

$$\phi = \frac{ds}{s^k} \wedge \psi + \frac{\theta}{s^{k-1}}. \tag{3.2}$$

When $k = 1$, *the restriction of* ψ *to the manifold* $\{s = 0\}$ *is d-closed. In this case, the form* $\psi_{|S}$ *is denoted* $\mathrm{res}_{\{s=0\}}(\phi)$.

Proof of Lemma 3.2. One can see first that

$$d(s^k \phi) = k s^{k-1} ds \wedge \phi$$

implies $ds \wedge (s^k \phi) = 0$; then, just using linear algebra (remember that one can suppose, up to a change of coordinates, that s is locally ζ_1, and then use a partition of unity), it is possible to write

$$d(s^k \phi) = ds \wedge \theta_1,$$

for some θ_1 which is C^∞ in U. Thus, the form $s^k \phi - (s\theta_1/k)$ is C^∞ in U and satisfies

$$ds \wedge (s^k \phi - \frac{s\theta_1}{k}) = \frac{s}{k}(kds \wedge s^{k-1}\phi - ds \wedge \theta_1) = 0.$$

So that, once again we have

$$s^k \phi - \frac{s\theta_1}{k} = ds \wedge \psi, \text{ for some } \psi \ C^\infty \text{ in } U.$$

Summarizing,

$$\phi = \frac{ds}{s^k} \wedge \psi + \frac{\theta_1}{ks^{k-1}} = \frac{ds}{s^k} \wedge \psi + \frac{\theta}{s^{k-1}},$$

which gives the desired conclusion. When $k = 1$, the same reasoning shows that $\psi_{|S}$ is a smooth d-closed form on S. In fact, from $s\phi = ds \wedge \psi + s\theta$, we conclude that

$$ds \wedge d\psi = d(s\phi) - ds \wedge d\theta - sd\theta = sd\theta,$$

hence,
$$ds_{|S} \wedge d\psi_{|S} = 0.$$

This clearly implies that $d\psi_{|S} = 0$. □

Remark 3.3. Note that the statement of Lemma 3.2 holds when U is replaced by an analytic manifold $\mathcal{X}$.

The construction of Res $_p[\phi]$, where $[\phi]$ is in $H^{p+1}(\mathcal{X} \backslash S)$ is as follows: first one chooses a semi-meromorphic representative of $[\phi]$, with poles in S. Then, one decreases the order of the pole by constructing in the same cohomology class (in $\mathcal{X} \backslash S$) some semi-meromorphic representative ϕ_0 with order 1, iterating Lemma 3.2. At this point, one defines Res $_p[\phi]$ as the cohomology class of $res_S \phi_0$ in $H^p(S)$.

Let us consider now the second case studied by Leray. Assume that all the S_j, for $j = 1, \ldots, M$, are $n-1$ dimensional submanifolds of $\mathcal{X}$ in general position. One can consider the inclusions

$$S_1 \cap \ldots \cap S_M \subset S_1 \cap \ldots \cap S_{M-1}$$
$$(S_1 \cap \ldots \cap S_{M-1}) \backslash S_M \subset S_1 \cap \ldots \cap S_{M-2},$$

$$S_1 \backslash (S_2 \cup \ldots \cup S_M) \subset \mathcal{X} \backslash (S_1 \cup \ldots \cup S_M)$$

and the corresponding sequence of Leray coboundary maps

$$H_p(S_1 \cap \ldots \cap S_M) \xrightarrow{\delta_p^{(M)}} H_{p+1}((S_1 \cap \ldots \cap S_{M-1}) \backslash S_M) \xrightarrow{\delta_{p+1}^{(M-1)}}$$

$$H_{p+2}((S_1 \cap \ldots \cap S_{M-2}) \backslash (S_{M-1} \cup S_M)) \to \ldots \xrightarrow{\delta_{p+M-1}^{(1)}} H_{p+M}(\mathcal{X} \backslash (S_1 \cup \ldots \cup S_M))$$

When ϕ is a d-closed $p+M$-form with simple poles on each S_j, which is smooth on $\mathcal{X} \backslash (S_1 \cup \ldots \cup S_M)$, one can consider the residue of ϕ, that is $res^{(1)}\phi$ corresponding to the inclusion $S_1 \backslash (S_2 \cup \ldots \cup S_M) \subset \mathcal{X} \backslash (S_1 \cup \ldots \cup S_M)$. This residue form is d-closed on $S_1 \backslash (S_2 \cup \ldots \cup S_M)$ and has simple poles on $S_1 \cap S_2, \ldots, S_1 \cap S_M$ (see Lemma 16.9 in [1], for a detailed proof of this last property.) So that one can define $res^{(2)}(res^{(1)}\phi)$, where the construction $res^{(2)}$ corresponds to the inclusion $(S_1 \cap S_2) \backslash (S_3 \cup \ldots \cup S_M) \subset S_1 \backslash (S_2 \cup \ldots \cup S_M)$. This is Leray's result for composite residues ([1, Theorem 16.10], [28].)

Theorem 3.4. Let $S_1, \ldots, S_M$ be submanifolds in general position in a complex analytic manifold $\mathcal{X}$. Let ϕ in $Z^{p+M}(\mathcal{X} \backslash (S_1 \cup \ldots \cup S_M))$, with simple poles on each S_j. Let σ in $Z_p(S_1 \cap \ldots \cap S_M)$ and γ in the homology class (in the homology group $H_p(\mathcal{X} \backslash (S_1 \cup \ldots \cup S_M))$ of $\delta_{p+M-1}^{(1)} \circ \ldots \circ \delta_p^{(M)}[\sigma]$; then

$$\int_\gamma \phi = (2\pi i)^M \int_\sigma res^{(M)} \circ \ldots \circ res^{(1)}(\phi). \tag{3.3}$$

Here again a procedure of division of differential forms provides a way to compute the iterated residue for a special type of differential forms that we will meet in the next sections. Consider, for example, the situation where the submanifolds S_j are defined globally by the functions s_j, and let us compute the left-hand side of equation (3.3) for a form with poles of order bigger than 1, but of of the following kind,

$$\phi = \frac{ds_1 \wedge \ldots ds_M \wedge \omega}{s_1^{r_1+1} \ldots s_M^{r_M+1}},$$

with $r_1, \ldots, r_M \in \mathbb{N}$ and $ds_1 \wedge \ldots \wedge ds_M \wedge d\omega = 0$, ω being a smooth p-form on $\mathcal{X}$. One can write

$$d\omega = ds_1 \wedge \omega_1 + \ldots ds_M \wedge \omega_M. \qquad (3.4)$$

for some $\omega_1, \ldots, \omega_M$ smooth on $\mathcal{X}$. This is again just linear algebra, since $S_1, \ldots, S_M$ are assumed to be in general position (see [28].) Differentiating (3.4), we obtain $ds_1 \wedge d\omega_1 + \ldots + ds_M \wedge d\omega_M = 0$ and $ds_1 \wedge \ldots \wedge ds_M \wedge d\omega_j = 0$, $j = 1, \ldots, M$. Therefore,

$$d\omega_j = ds_1 \wedge d\omega_{j,1} + \ldots + ds_M \wedge d\omega_{j,M}$$

where $\omega_{j,1}, \ldots, \omega_{j,M}$ are smooth. We can repeat this procedure and define $\omega_{l_1,l_2,\ldots,l_q}$, $l_i \in \{1, \ldots, M\}$ and $q \in \mathbb{N}$. Now , in $\mathcal{X} \setminus (S_1 \cup \ldots \cup S_M)$ one has

$$\frac{(-1)^k}{r_k+1} d\left(\frac{ds_{[k]} \wedge \omega}{s^{*(r+\underline{1})}}\right) + \frac{(-1)^{M-k}}{r_k+1} \frac{ds_{[k]} \wedge d\omega}{s^{*(r+\underline{1})}} = \frac{ds \wedge \omega}{s^{*(r+\underline{1})} \cdot s_k}$$

This identity shows that

$$\frac{ds \wedge \omega}{s^{*(r+\underline{1})} \cdot s_k} \cong \frac{(-1)^{M-k}}{r_k+1} \frac{ds_{[k]} \wedge d\omega}{s^{*(r+\underline{1})}} \cong \frac{1}{r_k+1} \frac{ds \wedge \omega_k}{s^{*(r+\underline{1})}},$$

where we denote by the symbol $\cong$ that the forms are cohomologous in the submanifold $\mathcal{X} \setminus (S_1 \cup \ldots \cup S_M)$. If we iterate this process, we obtain, for σ and γ as in Theorem 3.4, that

$$\int_\gamma \phi = \frac{(2\pi i)^M}{r_1! \ldots r_M!} \int_\gamma \frac{ds \wedge \omega_{1,\ldots,1,2,\ldots,2,\ldots,M,\ldots,M}}{s_1 \ldots s_M}.$$

Here $\omega_{1,\ldots,1,2,\ldots,2,\ldots,M,\ldots,M}$ means that 1 is repeated r_1 times, 2, r_2 times, etc. It follows from (3.3) that

$$\int_\gamma \frac{ds \wedge \omega_k}{s^{*(r+\underline{1})}} = (2\pi i)^M \int_\sigma \omega_{1,\ldots,1,2,\ldots,2,\ldots,M,\ldots,M}. \qquad (3.5)$$

One of the stumbling blocks in theory of Leray is the fact that $S_1, \ldots, S_M$ must be in general position. This is due to the fact that it is impossible in general to restrict distributions to submanifolds. This is precisely what the following hypotheses will allow us to do in the iterated procedure of Leray.

Example 1. Suppose that $f_1, \ldots, f_p$, $(p \le n)$ are p holomorphic functions in an open subset U of $\mathbb{C}^n$, such that any subfamily of $f_1, \ldots, f_p$ of cardinal k defines a submanifold of U of dimension $n - k$. Then, we say that $f_1, \ldots, f_p$ are in general position. If γ is an element in $Z_k(\{f_1 = \ldots f_p = 0\})$ and if ρ is a retract of a tubular neighborhood of $\{f_1 = \ldots f_p = 0\}$ onto $\{f_1 = \ldots f_p = 0\}$, then, $\rho^{-1}(\gamma)$ is a $(k + 2p)$-dimensional cell in U. Moreover, let us orient

$$\Gamma_{f,\gamma}(\epsilon) := \rho^{-1}(\gamma) \cap \{|f_1| = \epsilon_1\} \cap \ldots \cap \{|f_p| = \epsilon_p\},$$

with the orientation induced by that of $\rho^{-1}(\gamma)$ and the order $f_1, \ldots, f_p$, then $\Gamma_{f,\gamma}(\epsilon)$ is in the homology class in $U \setminus \{f_1 = \ldots = f_p = 0\}$ of $\delta^{(1)} \circ \ldots \circ \delta^{(p)}[\gamma]$, for all $\epsilon_1, \ldots, \epsilon_p$ sufficiently small . Therefore, for any $\phi \in Z^{k+p}(U)$,

$$\int_{\Gamma_{f,\gamma}(\epsilon)} \frac{\phi}{f_1 \ldots f_p} = (2\pi i)^p \int_\gamma res^{(p)} \circ \ldots \circ res^{(1)} \left(\frac{\phi}{f_1 \ldots f_p} \right). \qquad (3.6)$$

In particular, when $p = 1$ and $\phi \in H(U)$, it follows from Fubini's theorem that at any point were $f_1 = 0$ and $\partial f_1 / \partial \zeta_k \ne 0$,

$$res^{(1)}(\frac{\phi}{f_1} d\zeta) = (-1)^k \left(\left(\phi \frac{\partial f_1}{\partial \zeta_k} \right) |_{f_1 = 0} \right) d\zeta_{[k]}.$$

Example 2. When studying homogeneous polynomials of $n + 1$ variables, it is often convenient to consider both $\mathbb{P}^n(\mathbb{C})$ and $\mathbb{C}^{n+1}$ as subsets of $\mathbb{P}^{n+1}(\mathbb{C})$, respectively, the hyperplane at infinity and the affine space. For that purpose, recall that $\mathbb{P}^{n+1}(\mathbb{C})$ can be identified to the set of equivalence classes $[\zeta_0 : \zeta_1 : \ldots : \zeta_n : u] = [[z]]$, $z \in (\mathbb{C}^{n+2})^*$, where

$$[[z]] - [[z']] \iff \exists \lambda \in \mathbb{C}^* \text{ such that } z = \lambda z'.$$

The hyperplane Π_∞ of $\mathbb{P}^{n+1}(\mathbb{C})$, defined by $\{[[z]] : u = 0\}$, can be identified to $\mathbb{P}^n(\mathbb{C})$ and, its complement in $\mathbb{P}^{n+1}(\mathbb{C})$, can be identified to $\mathbb{C}^{n+1}$, i.e., we can write $\mathbb{C}^{n+1} = \{[[z]] : u = 1\}$.

Let Q be a homogeneous polynomial of $n + 1$ variables, $\mathcal{V}$ the projective algebraic variety in $\mathbb{P}^n(\mathbb{C})$ defined by $V := \{Q = 0\} \subset \mathbb{C}^{n+1}$. $\mathbb{P}^n(\mathbb{C}) \setminus \mathcal{V}$ can be viewed as a submanifold of $\mathbb{P}^{n+1}(\mathbb{C}) \setminus \bar{V}$, where $\bar{V}$ is the closure of V in $\mathbb{P}^{n+1}(\mathbb{C})$. From (2.1) we have $H_{n+2}(\mathbb{P}^{n+1}(\mathbb{C}) \setminus \bar{V}) = 0$. Since we have a retraction from $\mathbb{P}^{n+1}(\mathbb{C}) \setminus \bar{V}$ onto $\mathbb{P}^n(\mathbb{C}) \setminus \mathcal{V}$ (namely, $\tau : [\zeta_0 : \ldots : \zeta_n : u] \mapsto [\zeta_0 : \ldots : \zeta_n],$) we obtain [1,Corollary 0.20]

$$H_{n+1}(\mathbb{P}^{n+1}(\mathbb{C}) \setminus \bar{V}) = H_{n+1}(\mathbb{P}^n(\mathbb{C}) \setminus \mathcal{V}) = 0.$$

Therefore, it follows from the construction of the Leray coboundary sequence that the coboundary map δ is an isomorphism between $H_n(\mathbb{P}^n(\mathbb{C}) \setminus \mathcal{V})$ and $H_{n+1}((\mathbb{P}^{n+1}(\mathbb{C}) \setminus \bar{V}) \setminus (\mathbb{P}^n(\mathbb{C}) \setminus \mathcal{V})) = H_{n+1}(\mathbb{C}^{n+1} \setminus V)$.

On the other hand, any rational form ϕ of degree n in $\mathbb{P}^n(\mathbb{C})$ can be written in homogeneous coordinates in the form

$$\phi([\zeta]) = \frac{P(\zeta)}{Q(\zeta)} \sum_{k=0}^{n} (-1)^k \zeta_k d\zeta_{[k]},$$

where P, Q are homogeneous polynomials in $(\zeta_0, \ldots, \zeta_n)$ with $\deg(P) + n + 1 = \deg(Q)$. (For properties of differential forms in $\mathbb{P}^n(\mathbb{C})$, see, e.g., [6, §0].) Under these conditions, the form $\omega = P(d\zeta_0 \wedge \ldots \wedge d\zeta_n)/Q$ in $\mathbb{C}^{n+1}$ can be extended as a semi-meromorphic form $\tilde{\omega}$ to $\mathbb{P}^{n+1}(\mathbb{C})$, with a pole of order one on $\{u = 0\} \simeq \mathbb{P}^n(\mathbb{C})$. One way to do that is to replace the coordinates $\zeta_0, \ldots, \zeta_n$ in the expression of this differential form by $\zeta_0/u, \ldots, \zeta_n/u$ (remember that the form ω was a priori defined when $u = 1$); this gives immediately the following formula for $\tilde{\omega}$

$$\tilde{\omega}([\zeta : u]) = \frac{P(\zeta/u)}{Q(\zeta/u)} \bigwedge_{k=0}^{n} \left(\frac{d\zeta_k}{u} - \zeta_k \frac{du}{u^2}\right),$$

which can be rewritten as

$$\tilde{\omega}([\zeta : u]) = \frac{P(\zeta)}{Q(\zeta)} \left(\sum_{k=0}^{n} (-1)^k \zeta_k d\zeta_{[k]}\right) \wedge \frac{du}{u} + \theta,$$

where θ is smooth in $(\mathbb{P}^{n+1}(\mathbb{C}) \setminus \bar{V}) \setminus \{u = 0\}$; this division formula gives the value of

$$res\,\tilde{\omega} = \frac{P(\zeta)}{Q(\zeta)} \left(\sum_{k=0}^{n} (-1)^k \zeta_k d\zeta_{[k]}\right)$$

and we deduce from this computation the following interpretation of formula (3.1); namely, for σ in $Z_n(\mathbb{P}^n(\mathbb{C}) \setminus \mathcal{V})$, the formula becomes

$$2\pi i \int_\sigma \frac{P(\zeta)}{Q(\zeta)} \sum_{k=0}^{n} (-1)^k \zeta_k d\zeta_{[k]} = \int_{\delta\sigma} \frac{P(\zeta)}{Q(\zeta)} d\zeta_0 \wedge \ldots \wedge d\zeta_n. \qquad (3.7)$$

This provides a way to compute the integral of the rational form ϕ over a cycle σ in $\mathbb{P}^n(\mathbb{C}) \setminus \mathcal{V}$, where $\mathcal{V}$ denotes the set of poles of ϕ.

§2 Multiplication of principal values and residue currents

The fact that one can define $1/f$ as a distribution, when f is a holomorphic function in some subdomain U of $\mathbb{C}^n$, goes back to Schwartz [36]. The solution of the same problem when F is real analytic in some open subset of $\mathbb{R}^{2n}$ has been given by Lojasiewicz [27] and Hörmander [21]. The fundamental ingredient in their construction of such inverses (in the space of distributions) is the existence of what we usually call local Lojasiewicz inequalities, that is inequalities of the form

$$\|F(\zeta)\| \geq c \min(d(\zeta, \{F = 0\}))^k$$

for some exponent k, the inequality being understood as valid in a neighborhood of any zero ζ_0 of F, with $k = k(\zeta_0)$, $c = c(\zeta_0) > 0$.

Herrera and Lieberman, inspired by the work of Leray we just quoted, proved in [18] that one can define

$$PV[\frac{1}{f}] = \lim_{\epsilon \to 0}(\int_{|f|>\epsilon} \frac{\phi \, d\bar{\zeta} \wedge d\zeta}{f})$$

for any smooth function ϕ with compact support in $\mathbb{C}^n$. Such a distribution $PV[\frac{1}{f}]$, PV for *principal value* (see [13], [18]) is a solution T of $fT = 1$ and defines a $(0,0)$ current T_f in U. Let us compute $\bar{\partial} T_f$; one obtains, for any $(n, n-1)$ test form ψ,

$$< \bar{\partial} T_f, \psi > = - < T_f, \bar{\partial}\psi >$$

$$= - \lim_{\epsilon \to 0}(\int_{|f|>\epsilon} \frac{\bar{\partial}\psi}{f}) = - \lim_{\epsilon \to 0}(\int_{|f|>\epsilon} \bar{\partial}(\frac{\psi}{f})) = \lim_{\epsilon \to 0}(\int_{|f|=\epsilon} \frac{\psi}{f})$$

$$(3.8)$$

It is therefore clear that $\bar{\partial} T_f$ is a current supported by $\{f = 0\}$. If ψ is $\bar{\partial}$-closed near $\{f = 0\}$ and if f defines a manifold, then

$$< \bar{\partial} T_f, \psi > = 2\pi i \int_{\{f=0\}} res\left(\frac{\psi}{f}\right)$$

and the relation with the Leray residue is immediate. Note that this particular case (when f defines a submanifold), there is a different way to recover the same current, namely

$$< \bar{\partial} T_f, \psi > = \lim_{\lambda \to 0} \lambda \int_U |f|^{2(\lambda-1)} \overline{\partial f} \wedge \psi = \lim_{\lambda \to 0} \lambda \int_U \bar{\partial}(\frac{|f|^{2\lambda}}{f}) \wedge \psi.$$

It is natural to denote

$$\bar{\partial}\frac{1}{f} := \bar{\partial} T_f \quad [\frac{1}{f}] := T_f.$$

Remark 3.5. Note here that the existence of

$$\lim_{\epsilon \to 0} \Big(\int_{|f|=\epsilon} \frac{\psi}{f} \Big)$$

is far from being trivial when f does not define a submanifold in U; in fact, no argument based on the use of Lebesgue's dominated convergence theorem can in general be applied; let us take for example $n = 2$ and $f(\zeta) = \zeta_1^3 + \zeta_2^3$; let θ be a test function such that $(\partial\theta/\partial\bar{\zeta}_1)(0,0) \neq 0$; then, the integrals

$$\int_{|t^3+\eta^3|=\epsilon} \frac{\theta(t,\eta)}{t^3 + \eta^3} dt$$

cannot be uniformly bounded when (ϵ, η) belongs to any compact neighborhood of the origin. Let us choose (ϵ, η) in such a way that $\epsilon/\eta^3 \to 0$, and $\eta \to 0$. Then, by the change of variables $t = \eta u$,

$$\int_{|t^3+\eta^3|=\epsilon} \frac{\theta(t,\eta)}{t^3 + \eta^3} dt = \frac{1}{\eta^2} \int_{|u^3+1|=\epsilon/\eta^3} \frac{\theta(\eta u, \eta)}{u^3 + 1} du. \tag{3.9}$$

We now expand $\theta(v, \eta)$ as a Taylor series in $v, \bar{v}, \eta, \bar{\eta}$, near the origin. It is clear that the terms of order bigger or equal than 2 in the expansion remain bounded after integration and division by η^2. The remaining terms contribute to the right-hand side of (3.9) by linear combinations (with coefficients depending on $\eta, \bar{\eta}$) of the integrals

$$\int_{|u^3+1|=\epsilon/\eta^3} \frac{du}{u^3 + 1} \quad \text{and} \quad \int_{|u^3+1|=\epsilon/\eta^3} \frac{u\,du}{u^3 + 1}, \tag{3.10}$$

and the single term

$$(\partial\theta/\partial\bar{\zeta}_1)(0,0)\frac{\bar{\eta}}{\eta^2} \int_{|u^3+1|=\epsilon/\eta^3} \frac{\bar{u}\,du}{u^3 + 1}. \tag{3.11}$$

The two integrals in (3.10) can be computed by residues and they are zero because their rational integrands have no residue at ∞. From formula (1.4), the integral involved in (3.11) tends to

$$(2\pi i) \sum_{\alpha,\, \alpha^3=-1} \frac{\bar{\alpha}}{3\alpha^2} = -2\pi i \neq 0.$$

This implies that the integrals (3.9) cannot remain bounded in (ϵ, η) in any compact neighborhood of the origin.

Let us now introduce some useful tricks in order to perform multiplication of Principal Value distributions and residue currents. We have the following result, in the spirit of our Proposition 2.2.

Proposition 3.6. *Let $p \leq n$ and $f_1, \ldots, f_p, f_{p+1}, \ldots f_M$ be M holomorphic functions in an open set $U \subset \mathbb{C}^n$. Let ϕ be a $(n, n-p)$ test form with coefficients in $\mathcal{D}(U)$. The function of M variables $\lambda \mapsto J_f(\lambda, \phi)$ defined for $\operatorname{Re} \lambda_1 > 1, \ldots, \operatorname{Re} \lambda_M > 1$ by*

$$J_f(\lambda, \phi) = \frac{(-1)^{\frac{p(p-1)}{2}}}{(2\pi i)^p} \int \left(\prod_{k=p+1}^{M} \left(\frac{|f_k|^{2\lambda_k}}{f_k}\right) \right) \left(\bigwedge_{l=1}^{p} \bar{\partial} \left(\frac{|f_l|^{2\lambda_l}}{f_l}\right) \right) \wedge \phi. \qquad (3.12)$$

can be extended to a meromorphic function in $\mathbb{C}^M$. Moreover, there exists a polynomial P, product of linear forms in $\lambda_1, \ldots, \lambda_M$, with positive integer coefficients, such that the map $\lambda \mapsto P(\lambda) J_f(\lambda, \phi)$ is holomorphic in an orthant of $\mathbb{C}^M$ of the form $\{\operatorname{Re} \lambda_1 > -\delta, \ldots, \operatorname{Re} \lambda_M > -\delta\}$ for some $\delta > 0$.

For any $(t_1, \ldots, t_M)$ in $(\mathbb{R}^+)^M$, the function of one variable $\mu \mapsto J_f(\mu t, \phi)$ can be continued analytically to $\operatorname{Re} \mu > -\delta(t)$. The application $\phi \mapsto J_f(\mu t, \phi)_{\mu=0}$ is a $(0, p)$-current that will be denoted $R_t[f_1, \ldots, f_p; f_{p+1}, \ldots, f_M]$.

Remark 3.7. We have here a wide choice of multiplication procedures, apart from some very particular cases (among them, the case of complete intersections, with $p = M$, where all choices for t give the same current, as we will see in the next section). Many questions about them remain unsolved. For example, even in the case $M = p$, if the $f_1, \ldots, f_p$ do not define a complete intersection we do not know what is the relation between the currents defined above and the properties of the ideal $(f_1, \ldots, f_p)$. For instance, for $n = M = p = 2$, $f_1 = z_1$, $f_2 = z_1 z_2$, then $(f_1, f_2) = (f_1)$, it is a prime ideal, the associated variety is $z_1 = 0$, but the above currents acting on a form $\phi dz_1 \wedge dz_2$ yield

$$\frac{t_1 t_2}{(t_1 + t_2) t_2} \frac{\partial \phi}{\partial z_1}(0, 0)$$

which is supported by the origin, and this point plays no special role for the ideal (f_1, f_2).

Remark 3.8. In the Proposition 3.6, δ, $\delta(t)$, and P depend on ϕ, but may be kept constant as long as $\operatorname{supp}(\phi) \subseteq K$, $K \subset\subset U$.

Proof of Proposition 3.6. The method which consists in using analytic continuations (as distribution valued mappings) of functions of M complex variables of the form $\lambda \mapsto |f|^{*2\lambda}$ has been developed recently, due to its relation with the theory of $\mathcal{D}$-modules (the reader will find in [8], [9], [22], [29], [35], [36], much additional material necessary to consider all the remaining unansewered questions.) The proof of Proposition 3.6 is easy if one uses the powerful machinery of resolution of singularities, namely Hironaka's result [19] (for a new approach, see also [7].) Unfortunately, the price to pay for it is the lack of effectiveness in the proof; that is the main shortcoming of this approach. Let us prove our result in a direct way. Let $f := f_1 \cdots f_M$; the problem being a local one, we may restrict ourselves to a neighborhood, as small as we wish, of the origin, and assume $f(0) = 0$. It follows from Hironaka's theorem [19] that there exists a neighborhood of 0 in U, a complex n-dimensional manifold $\mathcal{X}$, a proper holomorphic map $\pi : \mathcal{X} \to U$ such that $\pi_{|\mathcal{X} \setminus \pi^* f^{-1}(0)}$ is a biholo-morphic map over $U \setminus f^{-1}(0)$, and such that for every $x_0 \in \mathcal{X}$, there are local coordinates w centered at x_0, such that, in a local chart around x_0,

$$\pi^* f(w) = u(w) w_1^{\alpha_1} \ldots w_n^{\alpha_n}$$

where $\alpha \in \mathbb{N}^n$ and u is a non vanishing holomorphic function. Because the ring of germs of holomorphic functions at the origin is a unique factorization domain, we have in a local chart around $x_0 \in \mathcal{X}$

$$\pi^* f_j(w) = u_j(w) w_1^{\alpha_{j,1}} \ldots w_n^{\alpha_{j,n}} = u_j w^{*\alpha_j} \, , j = 1, \ldots, M \, , \tag{3.13}$$

where $\alpha_j \in \mathbb{N}^n$ and u_j are non vanishing holomorphic functions.
For Re $\lambda_j \gg 0$, $j = 1, \ldots M$, and $\Sigma_\epsilon := \{|f| > \epsilon\}$,

$$J_f(\lambda, \phi) =$$

$$= \lim_{\epsilon \to 0} \int_{\mathrm{supp}(\phi) \cap \Sigma_\epsilon} \left(\prod_{k=p+1}^{M} \frac{|f_k|^{2\lambda_k}}{f_k} \right) \left(\bigwedge_{l=1}^{p} \bar{\partial} (\frac{|f_l|^{2\lambda_l}}{f_l}) \right) \wedge \phi$$

$$= \lim_{\epsilon \to 0} \int_{\pi^{-1}(\mathrm{supp}(\phi) \cap \Sigma_\epsilon)} \pi^* \left[\prod_{k=p+1}^{M} \frac{|f_k|^{2\lambda_k}}{f_k} \prod_{l=1}^{p} |f_l|^{2(\lambda_l - 1)} \right] \left(\bigwedge_{l=1}^{p} \lambda_l \overline{\partial \pi^* f_l} \right) \wedge \pi^* \phi$$

$$= \int_{\pi^{-1}(\mathrm{supp}(\phi))} \pi^* \left[\prod_{k=p+1}^{M} \frac{|f_k|^{2\lambda_k}}{f_k} \prod_{l=1}^{p} |f_l|^{2(\lambda_l - 1)} \right] \left(\bigwedge_{l=1}^{p} \lambda_l \overline{\partial \pi^* f_l} \right) \wedge \pi^* \phi.$$

We now use a partition of unity subordinated to the compact $\pi^{-1}(\mathrm{supp}(\phi))$ (remember π is proper) and J_f appears as a combination of terms of the form

$$j(\lambda) = \int |w|^{*2L(\lambda)} \frac{\lambda_1 \cdots \lambda_p}{w_1^{q_1} \cdots w_n^{q_n} \bar{w}_{j_1} \cdots \bar{w}_{j_p}} (\theta \xi \psi(\lambda, w)) d\bar{w} \wedge dw. \tag{3.14}$$

where the q_j are positive integers, $(j_1, \ldots, j_p)$ a p uplet of distinct elements of $\{1, \ldots, n\}$, θ a coefficient of $\pi^* \phi$, ξ some function from the partition of unity, $(\lambda, w) \mapsto \psi(\lambda, w)$ a function C^∞ in w, holomorphic in λ on $\mathbb{C} \times \mathrm{supp}(\xi)$, and finally $L_1, \ldots, L_n$ some linear forms with integer coefficients such that $L_j \neq 0$

if $q_j > 0$ or if $j \in \{j_1, \ldots, j_p\}$. Using Stokes's theorem (applicable when Re $\lambda_1 \gg 0, \ldots,$ Re $\lambda_M \gg 0$), one can rewrite (3.14) as

$$j(\lambda) = \frac{\lambda_1 \ldots \lambda_p}{L_{j_1}(\lambda) \ldots L_{j_p}(\lambda)} (-1)^p \times$$
$$\times \int |w|^{*2L(\lambda)} \frac{1}{w^{*q}} \frac{\partial^p}{\partial \bar{w}_{j_1} \ldots \partial \bar{w}_{j_p}} [\theta \xi \psi(\lambda, w)] d\bar{w} \wedge dw. \qquad (3.15)$$

We may now repeat an idea developed in Chapter 1 (see, for instance, the proof of Lemma 1.6) and obtain

$$j(\lambda) = (-1)^{p+|q|} \frac{\lambda_1 \ldots \lambda_p}{L_{j_1}(\lambda) \ldots L_{j_p}(\lambda)} \frac{1}{\prod\limits_{\substack{1 \leq k \leq n \\ q_k \geq 1}} \prod\limits_{j=1}^{q_k} (L_k(\lambda) + j)} \times$$
$$\times \int |w|^{*2L(\lambda)} \left(\frac{\bar{w}_1}{w_1}\right)^{q_1} \ldots \left(\frac{\bar{w}_n}{w_n}\right)^{q_n} \rho(\lambda, w) d\bar{w} \wedge dw, \qquad (3.16)$$

where

$$\rho(\lambda, w) = \frac{\partial^{p+|q|}}{\partial \bar{w}^q \partial \bar{w}_{i_1} \ldots \partial \bar{w}_{i_p}} [\theta \xi \psi(\lambda, w)].$$

The conclusion of the proposition follows at once from the fact that $J_f(\lambda, \phi)$ is a linear combination of expressions of the form (3.16). Note that using polar coordinates in (3.16), one can easily continue $\lambda \mapsto J_f(\lambda, \phi)$ as a meromorphic function in $\mathbb{C}^M$. $\qquad \square$

§3 The Dolbeault complex and the Grothendieck residue

As we have mentioned above, Proposition 3.6 is particularly interesting when $M = p$ and when $f_1, \ldots, f_p$ define a complete intersection in some open subset $U \subset \mathbb{C}^n$. This means that

$$\dim_{\zeta_0}\{f_1 = \ldots = f_p = 0\} = n - p$$

at any common zero ζ_0 of the f_j in U (here dim_ζ denotes the local dimension at some point ζ.) Note that this hypothesis is equivalent (see for example [16], p. 660) to the fact that for any ζ_0 in $\{f_1 = \ldots = f_p = 0\}$ and for any subset $\mathcal{J}$ of $\{1, \ldots, p\}$, the germs $\{(f_j)_{\zeta_0}, j \in \mathcal{J}\}$ at ζ_0 define a germ of analytic set of codimension exactly $n - \text{card}(\mathcal{J})$. The global equivalent of this assertion is false, as shown by the example $(\zeta_1(1-\zeta_3), \zeta_2(1-\zeta_3), \zeta_3)$ in $\mathbb{C}^3$: these functions define the origin, nevertheless, the first two ones define an hypersurface. In this section, as long as we are dealing with local results, we will most of the time make the assumption that the f_j define a complete intersection. Under these hypotheses, we have the following.

Proposition 3.9. *Let $f_1, \ldots, f_p$ be n holomorphic functions defining a complete intersection in a domain $U \subset \mathbb{C}^n$. Let ϕ be a $(n, n-p)$ test form with coefficients in $\mathcal{D}(U)$ such that $\bar\partial\phi = 0$ in a neighborhood of the set $\{f_1 = \ldots = f_n = 0\}$. There exists $A > 0$ and a subset Δ of measure 0 in $(\mathbb{R}^+)^p \times (\mathbb{R}^+)^p$ such that, for any $(\epsilon, \epsilon') \in ((\mathbb{R}^+)^p \times (\mathbb{R}^+)^p) \backslash \Delta$, with the property that $\max(\|\epsilon\|, \|\epsilon'\|) \leq A$,*

$$\int_{\substack{|f_j|=\epsilon_j \\ 1 \leq j \leq p}} \frac{\phi}{f_1 \cdots f_p} = \int_{\substack{|f_j|=\epsilon'_j \\ 1 \leq j \leq p}} \frac{\phi}{f_1 \cdots f_p}. \tag{3.17}$$

Remark 3.10. We have to make precise the orientation we consider on the set $\{|f_1| = \epsilon_1, \ldots, |f_p| = \epsilon_p\}$, which, as we will see later, remains a submanifold for generic values of ϵ. This can be done just by prescribing the rules of integration of differential forms on such a set; here we impose the following: if $(\chi_\nu)_{\nu \geq 0}$ is a sequence of smooth regularizations of the characteristic function of $[1, +\infty[$, then, for any test form ψ of degree $2n - p$,

$$\int_{\substack{|f_i|=\epsilon_i \\ 1 \leq i \leq p}} \psi := (-1)^{\frac{p(p-1)}{2}} \lim_{\nu \to \infty} \int d(\chi_{1,\nu}^{\epsilon_1}) \wedge \cdots \wedge d(\chi_{p,\nu}^{\epsilon_p}) \wedge \psi \tag{3.18}$$

where $\chi_{j,\nu}^{\epsilon_j}$ is defined by $\chi_{j,\nu}^{\epsilon_j} = \chi_\nu(|f_j|^2 \epsilon_j^{-2})$. Henceforth we will denote the set $\{|f_1| = \epsilon_1, \ldots, |f_p| = \epsilon_p\}$ as $\Gamma_f(\epsilon_1, \ldots, \epsilon_p)$, or more simply by $\Gamma(\epsilon)$; note that the ordering of ϵ is important.

Proof of the Proposition 3.9. We choose $A > 0$ so that in $\text{supp}(\phi) \cap \{\|f\| \leq A\}$ one has $\bar\partial\phi \equiv 0$. Such a choice of A is possible because $\text{supp}(\phi)$ is a compact subset in U. We now use Sard's theorem (see Chapter 2, reference [19]) to construct a set Δ of measure 0 in $(\mathbb{R}^+)^{2p}$ such that if $\delta = (\epsilon, \epsilon') \notin \Delta$, for any set $\mathcal{E} \subset \{\epsilon_1, \ldots, \epsilon_p, \epsilon'_1, \ldots, \epsilon'_p\}$ and any set $\mathcal{J} \subset \{1, \ldots, p\}$ with $\text{card}(\mathcal{E}) = \text{card}(\mathcal{J}) = l$, then $(\delta_{j_1}, \ldots, \delta_{j_l})$ is not a critical value for $(|f_{j_1}|^2, \ldots, |f_{j_l}|^2)$. Choose now ϵ, ϵ' with $\max(\|\epsilon\|, \|\epsilon'\|) \leq A$ and $(\epsilon, \epsilon') \notin \Delta$. Using local coordinates near each point of $\{|f_1| = \epsilon_1, \ldots, |f_p| = \epsilon_p\}$ (which is smooth), we have

$$\int_{\Gamma(\epsilon)} \frac{\phi}{f_1 \cdots f_p} = (-1)^{\frac{p(p-1)}{2}} \lim_{\nu \mapsto \infty} \int_{|f_1|=\epsilon_1} \bar\partial\chi_{2,\nu}^{\epsilon_2} \wedge \cdots \wedge \bar\partial\chi_{p,\nu}^{\epsilon_p} \wedge \frac{\phi}{f_1 \cdots f_p}$$

$$\int_{\Gamma(\epsilon'_1, \epsilon_2, \ldots, \epsilon_p)} \frac{\phi}{f_1 \cdots f_p} = (-1)^{\frac{p(p-1)}{2}} \lim_{\nu \mapsto \infty} \int_{|f_1|=\epsilon'_1} \bar\partial\chi_{2,\nu}^{\epsilon_2} \wedge \ldots \wedge \bar\partial\chi_{p,\nu}^{\epsilon_p} \wedge \frac{\phi}{f_1 \cdots f_p}$$

The differential form

$$\psi_\nu^{(\epsilon)} := \frac{\bar\partial\chi_{2,\nu}^{\epsilon_2}}{f_2} \wedge \ldots \wedge \frac{\bar\partial\chi_{p,\nu}^{\epsilon_p}}{f_p} \wedge \phi$$

is $\bar\partial$-closed in $\mathbb{C}^n$ and has no singularities. Therefore, the integral

$$\int_{\partial\{\epsilon_1'\leq|f_1|\leq\epsilon_1\}}\psi_\nu^\epsilon$$

is zero because of Stokes's theorem. We conclude then that

$$\int_{\Gamma(\epsilon)}\frac{\phi}{f_1\cdots f_p}=\int_{\Gamma(\epsilon_1',\epsilon_2,\ldots,\epsilon_p)}\frac{\phi}{f_1\cdots f_p}$$

We can now repeat this argument, using f_2 instead of f_1; the proposition follows by iteration. $\square$

In order to introduce the concept of Grothendieck residue, we need to recall a few facts about the Dolbeault complex. If $\mathcal{X}$ is a n-dimensional complex manifold and $\mathcal{S}$ a closed subset of $\mathcal{X}$, we introduce the collection of vector spaces $\Lambda_c^{k,l}(\mathcal{X})$, (respectively, $\Lambda_c^{k,l}(\mathcal{X}\setminus\mathcal{S})$) of (k,l) differential forms ($k\leq n, l\leq n$) with coefficients in $\mathcal{D}(\mathcal{X})$ (respectively in $\mathcal{D}(\mathcal{X}\setminus\mathcal{S})$); then we define

$$\Lambda^{k,l}[\mathcal{S}]:=\frac{\Lambda_c^{k,l}(\mathcal{X})}{\Lambda_c^{k,l}(\mathcal{X}\setminus\mathcal{S})},$$

which is the vector space of germs of such forms on $\mathcal{S}$. The Dolbeault complex can be represented as follows

where i is the natural imbedding of of $\Lambda_c^{(\cdot,\cdot)}(\mathcal{X}\setminus\mathcal{S})$ into $\Lambda_c^{(\cdot,\cdot)}(\mathcal{X})$ and p the quotient map from $\Lambda_c^{(\cdot,\cdot)}(\mathcal{X})$ into $\Lambda^{(\cdot,\cdot)}[\mathcal{S}]$. We associate to the last row of the diagram cohomology groups $H^{(\cdot,\cdot)}[\mathcal{S}]$. To any $(n-k,n-l)$ $\bar\partial$-closed current T

on $\mathcal{X} \setminus \mathcal{S}$ $(0 \leq k \leq n, 1 \leq l \leq n)$, one can associate an element in the dual of $H^{k,l-1}[\mathcal{S}]$, denoted as Res $[T]$; this linear application is defined by

$$\text{Res}\,[T](h) = T(\bar{\partial}\phi)$$

where ϕ is any element in $\Lambda_c^{k,l-1}(\mathcal{X})$ which satisfies $\bar{\partial}\phi = 0$ near $\mathcal{S}$ and represents the equivalence class h in $H^{k,l-1}[\mathcal{S}]$. It is clear that the value $T(\bar{\partial}\phi)$ does not depend of the choice of the representative ϕ; this follows from the fact that T is closed outside $\mathcal{S}$. We are now ready to define the Grothendieck residue and to show that, as a residue in the sense of Dolbeault (that means, in the cohomological sense, just as above), it can be obtained as Res $[T]$ for a current T that we will make explicit below. Let us first define this new object.

Definition 3.11. *Let $f_1, \ldots, f_p$ be p holomorphic functions in an open set $U \subset \mathbb{C}^n$ defining a complete intersection in U. The Grothendieck residue associated to $(f_1, \ldots, f_p)$, is denoted by $\bar{\partial}(1/f_1) \wedge \cdots \wedge \bar{\partial}(1/f_p)$, and it is the element of the dual of the vector space $H^{n,n-p}[\{f_1 = \ldots = f_p = 0\}]$ defined by*

$$< \bar{\partial}\frac{1}{f_1} \wedge \ldots \wedge \bar{\partial}\frac{1}{f_p}, h >= \lim_{\epsilon \to 0} \frac{1}{(2\pi i)^p} \int_{\Gamma_f(\epsilon)} \frac{\phi_h}{f_1 \ldots f_p},$$

where ϕ_h is any element in $\Lambda_c^{n,n-p}(U)$ such that $\bar{\partial}\phi_h = 0$ in a neighborhood of the variety $\{f_1 = \ldots = f_p = 0\}$), which is a representative of h in $H^{n,n-p}[\{f_1 = \ldots = f_p = 0\}]$.

This definition cannot be completely understood without the help of the next proposition.

Proposition 3.12. *Let $f_1, \ldots, f_p$ be as above and ϕ be a smooth $(n, n - p)$ form with coefficients in $\mathcal{D}(U)$, which is $\bar{\partial}$-closed near $\{f_1 = \ldots = f_p = 0\}$; then, for almost any $\epsilon \in (\mathbb{R}^+)^p$ such that ϕ remains $\bar{\partial}$-closed in a neighborhood of $\{|f_1| \leq \epsilon_1, \ldots, |f_p| \leq \epsilon_p\}$, we have*

$$\frac{1}{(2\pi i)^p} \int_{\Gamma_f(\epsilon)} \frac{\phi}{f_1 \ldots f_p} = \frac{(-1)^{\frac{p(p-1)}{2}} (p-1)!}{(2\pi i)^p} \int_{\substack{|f_j| \leq \epsilon_j \\ 1 \leq j \leq p}} \frac{\sum_{k=1}^{p} (-1)^{k-1} \bar{f}_k d\bar{f}_{[k]}}{\|f\|^{2p}} \wedge \phi$$

(3.19)

and

$$\frac{1}{(2\pi i)^p} \int_{\Gamma_f(\epsilon)} \frac{\phi}{f_1 \ldots f_p} = \frac{(-1)^{\frac{(p-1)(p+2)}{2}} (p-1)!}{(2\pi i)^p} \int_{\substack{|f_j| \leq \epsilon_j \\ 1 \leq j \leq p}} \frac{\sum_{k=1}^{p} (-1)^{k-1} \bar{f}_k d\bar{f}_{[k]}}{\|f\|^{2p}} \wedge \bar{\partial}\phi.$$

(3.20)

Note that (3.20) is an immediate consequence of (3.19) and of Stokes's formula; this follows from the fact that the differential form

$$\zeta \mapsto \Omega(f, \zeta) := \frac{1}{\|f\|^{2p}} \sum_{k=1}^{p} (-1)^{k-1} \bar{f}_k d\bar{f}_{[k]}$$

is $\bar\partial$-closed outside $\{f_1 = \ldots = f_p = 0\}$); up to a constant, this differential form, considered as a current outside this set, provides the current T from which the Grothendieck residue can be derived (as a cohomological residue in the sense of Dolbeault.)

We are going to give here a proof of this proposition that is slightly different from the usual one (compare for example [16], p. 653-654.) The reason for that is that we have in mind to extend our results to general test forms and see if one can make operate the residue we just defined on the whole class of test forms, so that $\mathrm{Res}\,[T]$ could really be considered as a $(0, p)$ current in U with support in $\{f_1 = \ldots = f_p = 0\}$.

We will use a classical lemma from differential geometry.

Lemma 3.13. *Let* $f_1, \ldots, f_p$ *be* p *holomorphic functions defining a complete intersection in a domain* $U \subset \mathbb{C}^n$. *Let* ϕ *be a smooth* $(n, n - p)$ *form with coefficients in* $\mathcal{D}(U)$. *Let* $s \mapsto I_f(\phi, s)$ *be the function defined at all non critical values of* $(|f_1|^2, \ldots, |f_p|^2)$ *by*

$$I_f(\phi, s_1, \ldots, s_p) = \frac{1}{(2\pi i)^p} \int\limits_{\substack{|f_j|^2 = s_j \\ 1 \le j \le p}} \frac{\phi}{f_1 \cdots f_p}.$$

For any function continuous function g *on* $(\mathbb{R}^+)^p$, *and for any real numbers* $0 < \eta_1 < \xi_1, \ldots, 0 < \eta_p < \xi_p$, *we have*

$$\int\limits_{\eta_1}^{\xi_1} \cdots \int\limits_{\eta_p}^{\xi_p} g(s) I_f(\phi, s) dm(s) = \frac{(-1)^{\frac{p(p-1)}{2}}}{(2\pi i)^p} \int\limits_{\substack{\eta_j \le |f_j|^2 \le \xi_j \\ 1 \le j \le p}} g(|f_1|^2, \ldots, |f_p|^2) \bar\partial f \wedge \phi.$$

$$(3.21)$$

Proof of the Lemma 3.13. From the Coarea Formula (see [14], Theorem 3.2.11, p. 248), it follows that the function defined almost everywhere as the $2n - p$ dimensional Lebesgue measure m_{2n-p} of $\mathrm{supp}(\phi) \cap \Gamma_f(\epsilon)$ is locally integrable (as a function of ϵ in $([0, \infty[)^p$. In fact, one has the estimate

$$\int m_{2n-p}(\mathrm{supp}(\phi) \cap \Gamma_f(\epsilon)) dm(\epsilon) \le C(\max_{1 \le j \le p} \|\mathrm{grad}(f_j)\|_{\mathrm{supp}(\phi)})^p \,\mathrm{vol}(\mathrm{supp}(\phi))$$

$$(3.22)$$

for some absolute constant $C = C(n, p)$. Therefore, Lebesgue's dominated convergence theorem implies that it is sufficient to prove (3.21) when $[\eta_1, \xi_1] \times$

$\ldots \times [\eta_p, \xi_p]$ is a small neigborhood of a noncritical value s_0 of $(|f_1|^2, \ldots, |f_p|^2)$. In that case, we already know that the sequence $s \mapsto I_f^{(\nu)}(\phi, s)$, $\nu \in \mathbb{N}$ defined as

$$I_f^{(\nu)}(\phi, s) = \frac{(-1)^{\frac{p(p-1)}{2}}}{(2\pi i)^p} \int \bar{\partial}(\chi_\nu(\frac{|f_1|^2}{s_1})) \wedge \ldots \wedge \bar{\partial}(\chi_\nu(\frac{|f_p|^2}{s_p})) \wedge \frac{\phi}{f_1 \ldots f_p}$$

converges uniformly on $[\eta_1, \xi_1] \times \ldots \times [\eta_p, \xi_p]$ to the function $s \mapsto I_f(\phi, s)$. Therefore,

$$\lim_{\nu \to \infty} \left(\int_{\eta_1}^{\xi_1} \ldots \int_{\eta_p}^{\xi_p} g(s_1, \ldots, s_p) I_f^{(\nu)}(\phi, s) dm(s) \right) =$$

$$= \int \left[\int_{\eta_1}^{\xi_1} \ldots \int_{\eta_p}^{\xi_p} g(s) \chi_\nu'(\frac{|f_1|^2}{s_1}) \cdots \chi_\nu'(\frac{|f_p|^2}{s_p}) dm(s) \right] \overline{\partial f} \wedge \phi =$$

$$= \int \Theta_\nu(\zeta) \overline{\partial f} \wedge \phi, \tag{3.23}$$

where $\Theta_\nu(\zeta)$ represents the expression between brackets in the previous line. Let us fix $\zeta = \zeta_0$, which is not a zero of any othe f_j, and observe that, as a sequence of distributions in the variable $t \in \mathbb{R}$, the sequence of functions $t \mapsto |f_j(\zeta_0)|^2 \chi_\nu'(|f_j(\zeta_0)|^2 t)$, $\nu \in \mathbb{N}$, converges to the Dirac mass at $|f_j(\zeta_0)|^{-2}$. Performing integration by parts, one can see that the sequence Θ_ν inside the right-handside integrand in (3.23) converges pointwise to

$$g(|f_1|^2, \ldots, |f_p|^2) \chi_{[\eta_1, \xi_1] \times \ldots \times [\eta_p, \xi_p]}(|f_1|^2, \ldots, |f_p|^2)$$

and remains bounded. Once again we apply Lebesgue's theorem and the formula (3.21) is proved. This ends the proof of Lemma 3.13.   □

Proof of Proposition 3.12. As ϕ is $\bar{\partial}$ closed near $\{|f_1| \leq \epsilon_1, \ldots, |f_p| \leq \epsilon_p\}$, the function $s \mapsto I_f(\phi, s)$ is almost everywhere constant on $]0, \epsilon_1^2[\times \ldots \times]0, \epsilon_p^2[$ (this follows from Proposition 3.9.) We use formula (3.21) with

$$g_\tau(s) = \frac{\tau}{(s_1 + \ldots + s_p + \tau)^{p+1}}, \quad \tau > 0$$

From a Tauberian argument we have already used in the case of one variable (see Proposition 1.8), we get

$$
\lim_{\tau \to 0} p! \int_0^{\epsilon_1^2} \cdots \int_0^{\epsilon_p^2} \frac{\tau I_f(\phi, s)}{(s_1 + \ldots + s_p + \tau)^{p+1}} \, dm(s) = < \bar{\partial} \frac{1}{f_1} \wedge \cdots \wedge \bar{\partial} \frac{1}{f_p}, [\phi] >
$$

$$
= (-1)^{\frac{p(p-1)}{2}} \frac{p!}{(2\pi i)^p} \lim_{\tau \to 0} \left(\int_{\substack{|f_j| \leq \epsilon_j \\ 1 \leq j \leq p}} \frac{\tau \overline{\partial} f}{(\|f\|^2 + \tau)^{p+1}} \wedge \phi \right)
$$

$$
= (-1)^{\frac{p(p-1)}{2}} \frac{(p-1)!}{(2\pi i)^p} \lim_{\tau \to 0} \int_{\substack{|f_j| \leq \epsilon_j \\ 1 \leq j \leq p}} \bar{\partial} \left[\frac{\sum_{k=1}^{p} (-1)^{k-1} \bar{f}_k d\bar{f}_{[k]}}{(\|f\|^2 + \tau)^p} \right] \wedge \phi.
$$

This integral can be rewritten using Stokes's formula when the boundary of $\{|f_1| \leq \epsilon_1, \ldots, |f_p| \leq \epsilon_p\}$ is piecewise smooth. In this way one obtains formula (3.19) and reaches the conclusion of the proof. □

As a consequence of the same principles, one can obtain the Grothendieck residue using the multidimensional Mellin Transform of the map $s \mapsto I_f(\phi, s)$.

Proposition 3.14. Let $f_1, \ldots, f_p$ be p holomorphic functions defining a complete intersection in a domain $U \subset \mathbb{C}^n$. For any $(n, n-p)$ test form ϕ which is $\bar{\partial}$-closed in a neighborhood of $\{f_1 = \ldots = f_p = 0\}$ and for any $t \in (\mathbb{R}^+)^p$ we have

$$
R_t[f_1, \ldots, f_p] = < \bar{\partial} \frac{1}{f_1} \wedge \cdots \wedge \bar{\partial} \frac{1}{f_p}, [\phi] > .
$$

Remark 3.15. Let us describe the relation between the function $\lambda \mapsto J_f(\lambda, \phi)$ introduced in (3.12) and the multidimensional Mellin Transform of the function $s \mapsto I_f(\phi, s)$. Introduce complex parameters $\lambda_1, \ldots, \lambda_p$ with $\mathrm{Re}\, \lambda_1 > 1, \ldots,$ $\mathrm{Re}\, \lambda_p > 1$, and deduce from Lemma 3.13 that, for any $\xi, \eta \subset (\mathbb{R}^+)^p$, such that $0 < \eta_1 < \xi_1, \ldots, 0 < \eta_p < \xi_p,$

$$
\lambda_1 \ldots \lambda_p \int_{\substack{\eta_j \leq s_j \leq \xi_j \\ 1 \leq j \leq p}} s^{*(\lambda-1)} I_f \, dm(s) = \frac{(-1)^{\frac{p(p-1)}{2}}}{(2\pi i)^p} \int_{\substack{\eta_j \leq |f_j|^2 \leq \xi_j \\ 1 \leq j \leq p}} \left(\bigwedge_{l=1}^{p} \bar{\partial}[\frac{|f_l|^{2\lambda_l}}{f_l}] \right) \wedge \phi.
$$

On the other hand, in terms of the notations preceding the inequality (3.22), we have that for almost all s,

$$
|I_f(\phi, s)| \leq \frac{1}{\sqrt{s_1} \cdots \sqrt{s_p}} \|\phi\| m_{n-2p}(\mathrm{supp}(\phi) \cap \{|f_1| = \sqrt{s_1}, \ldots, |f_p| = \sqrt{s_p}\}).
$$

It follows from (3.22) that the function

$$\lambda \mapsto \lambda_1 \ldots \lambda_p \int_{[0,\infty[^p} s^{*(\lambda-\underline{1})} I_f(\phi, s) dm(s)$$

is defined and holomorphic in the region Re $\lambda_1 > 1, \ldots,$ Re $\lambda_p > 1$. Therefore, the holomorphic function $\lambda \mapsto J_f(\lambda, \phi)$ is the Mellin Transform of the function $s \mapsto I_f(\phi, s)$. The fact that I_f (as a function of s) is constant almost everywhere in $]0, \sigma_1[\times \ldots]0, \sigma_p[$ for some $\sigma_j > 0$ does not imply as, in the one dimensional case, that the function $\lambda \mapsto J_f(\lambda, \phi)$ can be continued as an holomorphic function up to the origin. Nevertheless, what we are going to prove here is that all the values $R_t[f_1, \ldots, f_p]$ coincide with this constant.

Proof of the Proposition 3.14. For $t \in (\mathbb{R}^+)^p$ and Re $\mu \gg 0$ we can use formula (3.22) to obtain

$$\mu \int_{(\mathbb{R}^+)^p} s^{*\mu t} \frac{I_f(\phi, s)}{(s_1 + \cdots + s_p)^p} dm(s) = \frac{(-1)^{\frac{p(p-1)}{2}}}{(2\pi i)^p} \mu \int |f|^{*2\mu t} \frac{\overline{\partial f}}{\|f\|^{2p}} \wedge \phi \quad (3.24)$$

Since $s \mapsto I_f(\phi, s)$ is almost everywhere constant near the origin, let us say in $[0, A]^p$, the left-hand side of (3.24) defines a function of μ which is equal, up to an entire function vanishing at the origin, to

$$I_f(\phi, \underline{0}) A^{t_1 + \cdots + t_p} \mu \frac{\prod_{k=1}^{p} \Gamma(t_k \mu + 1)}{(t_1 + \cdots + t_p)\Gamma((t_1 + \ldots + t_p)\mu + p)}. \quad (3.25)$$

(see formula 2, p. 622 in [15].) Therefore, the function of the complex variable μ defined for Re $\mu \gg 0$ by

$$\mu \mapsto J_f(t, \mu; \phi) = \mu \int |f|^{*2\mu t} \frac{\overline{\partial f}}{\|f\|^{2p}} \wedge \phi$$

has an analytic continuation to the complex plane as a meromorphic function; moreover, its analytic continuation is holomorphic at the origin and satifies

$$J_f(t, 0; \phi) = \frac{I_f(\phi, \underline{0})}{(p-1)!(t_1 + \cdots + t_p)}. \quad (3.26)$$

Note that the poles of this meromorphic continuation are completely explicit from (3.25). We are now going to transform the expression $J(t, \mu; \phi)$ using Stokes's formula. For $2 \le j \le p$, we have, always assuming that Re $\mu \gg 0$,

$$(-1)^j \bar{\partial} \left(\frac{1}{f_j} |f|^{*2\mu t} \frac{\overline{\partial f_{[j]}}}{\|f\|^{2(p-1)}} \right) =$$

$$- \mu t_j |f_j|^{2(\mu t_j - 1)} |f_1|^{2\mu t_1} \cdots \widehat{|f_j|}^{2\mu t_j} \cdots |f_p|^{2\mu t_p} \frac{\overline{\partial f}}{\|f\|^{2p}} + (p-1)|f|^{*2\mu t} \frac{\overline{\partial f}}{\|f\|^{2p}}$$

$$(3.27)$$

Therefore,

$$
\begin{aligned}
\int |f|^{*2\mu t}\frac{\overline{\partial f}\wedge\phi}{\|f\|^{2p}} &= \frac{\mu t_j}{p-1}\int |f_j|^{2(\mu t_j-1)}|f_1|^{2t_1\mu}\cdots\widehat{|f_j|}^{2t_j\mu}|f_p|^{2t_p\mu}\frac{\overline{\partial f}}{\|f\|^{2p}} \\
&\quad + \frac{(-1)^j}{p-1}\int \bar\partial\left[\frac{1}{f_j}|f|^{*2\mu t}\frac{\overline{\partial f_{[j]}}}{\|f\|^{2(p-1)}}\right]\wedge\phi \\
&= \frac{t_j\mu}{p-1}\int |f_j|^{2(t_j\mu-1)}|f_1|^{2t_1\mu}\cdots\widehat{|f_j|}^{2t_j\mu}|f_p|^{2t_p\mu}\frac{\overline{\partial f}}{\|f\|^{2p}} \\
&\quad + \frac{(-1)^{j+p-1}}{p-1}\int \frac{1}{f_j}|f|^{*2\mu t}\frac{\overline{\partial f_{[j]}}}{\|f\|^{2(p-1)}}\wedge\bar\partial\phi .
\end{aligned}
$$

$$(3.28)$$

If $\bar\partial\phi$ vanishes on a neighborhood of the set of common zeros V (or $\bar\partial\phi$ vanishes to a sufficiently high order on V), the form $\|f\|^{2(p-1)}\overline{\partial f_{[j]}}\wedge\bar\partial\phi$ is a sufficiently smooth form with compact support. The second integral in the right-hand side of (3.28) defines, by Proposition 3.6, a meromorphic function of μ, regular at the origin. We repeat this procedure and conclude that, for any $k\in\{1,\dots,p\}$,

$$
\int |f|^{*2\mu t}\frac{\overline{\partial f}\wedge\phi}{\|f\|^{2p}} = \frac{t_1\dots\hat t_k\dots t_p}{(p-1)!}\mu^{p-1}\int |f|^{*2(\mu t-1)}|f_k|^2\frac{\overline{\partial f}\wedge\phi}{\|f\|^{2p}} + R(\mu),
$$

where R is a meromorphic function, which is holomorphic at $\mu=0$. We have then

$$
\mu(t_1+\cdots+t_p)\int |f|^{*2\mu t}\frac{\overline{\partial f}}{\|f\|^{2p}}\wedge\phi = \frac{1}{(p-1)!}\int \left(\bigwedge_{l=1}^{p}\bar\partial[\frac{|f_l|^{2\mu t_l}}{f_l}]\right)\wedge\phi + R_1(\mu),
$$

with R_1 also meromorphic and regular at the origin. From (3.25) and the definition of $R_t[f_1,\dots,f_p]$, we obtain the formula we were looking for. □

Let us use the same ideas to obtain a result that shall be useful to study division formulas.

Proposition 3.16. *Let $f_1,\dots,f_p$ be p holomorphic functions defining a complete intersection in a domain $U\subset\mathbb{C}^n$. For any $(n,n-p)$ test form ϕ, $\bar\partial$-closed in a neighborhood of $\{f_1=\dots=f_p=0\}$, any $t\in(\mathbb{R}^+)^p$, and any $\beta\in\mathbb{N}^n$, the function of one complex variable μ, defined for $\mathrm{Re}\,\mu\gg 0$ by*

$$
\mu\mapsto \mathcal{J}_f^\beta(t,\mu;\phi) = \mu\int |f|^{*2\mu t}\frac{\bar f^{*\beta}\overline{\partial f}}{\|f\|^{2(p+|\beta|)}}\wedge\phi ,
$$

has a meromorphic continuation to the whole complex plane, which is regular at the origin, and such that

$$
\mathcal{J}_f^\beta(t,0;\phi) = \frac{(-1)^{\frac{p(p-1)}{2}}(2\pi i)^p\beta!}{(t_1+\dots+t_p)(p+|\beta|-1)!}<\bar\partial\frac{1}{f_1^{\beta_1+1}}\wedge\cdots\wedge\bar\partial\frac{1}{f_p^{\beta_p+1}},[\phi]>.
$$

$$(3.29)$$

Proof. For Re $\mu \gg 0$, we rewrite J_f^β as

$$J_f^\beta(t,\mu;\phi) = \frac{\mu}{(\beta_1+1)\cdots(\beta_p+1)} \int |f|^{*2\mu t} \frac{\overline{\partial f^{\beta+\underline{1}}}}{\|f\|^{2(p+|\beta|)}} \wedge \phi.$$

Let us denote by $\tilde{\Gamma}(s;\beta)$ the cycle $\{|f_j|^{2(\beta_j+1)} = s_j,\ 1 \le j \le p\}$ and

$$\theta_\beta(s) := \int_{\tilde{\Gamma}(s;\beta)} \frac{\phi}{f_1^{\beta_1+1}\cdots f_p^{\beta_p+1}}.$$

Using this notation and formula (3.21), we deduce that

$$J_f^\beta(t,\mu;\phi) = \frac{\mu}{(\beta_1+1)\cdots(\beta_p+1)} \int_{(\mathbb{R}^+)^p} \theta_\beta(s) \frac{\prod_{j=1}^p s_j^{\frac{\mu t_j}{\beta_j+1}}}{\left(\sum_{j=1}^p s_j^{\frac{1}{\beta_j+1}}\right)^{p+|\beta|}} \, dm(s) \tag{3.30}$$

Since $\theta_\beta(s)$ is almost everywhere constant near the origin, there is a constant $A > 0$ such that the right-hand side of (3.30) equals, up to un entire function of μ vanishing at the origin,

$$\theta_\beta(0) A^{\mu \sum_{j=1}^p t_j/(\beta_j+1)} \frac{\prod_{j=1}^p \Gamma(\mu t_j + \beta_j + 1)}{(t_1 + \cdots + t_p)\Gamma(\mu(t_1 + \cdots + t_p) + p + |\beta|)}$$

(see again [37, Formula 2, p. 622].) Letting $\mu = 0$, the identity (3.29) follows immediately.   □

It is convenient to write

$$< \bar{\partial}\frac{1}{f^\alpha}, [\phi] > := < \bar{\partial}\frac{1}{f_1^{\alpha_1}} \wedge \cdots \wedge \bar{\partial}\frac{1}{f_p^{\alpha_p}}, [\phi] > . \tag{3.31}$$

Remark 3.17. Proposition 3.16, combined with Stokes's theorem, provides another way to compute

$$< \bar{\partial}\frac{1}{f^{\beta+\underline{1}}}, [\phi] > .$$

Namely, if ϵ is such that $\Delta_\epsilon := \{|f_1| \le \epsilon_1, \ldots, |f_p| \le \epsilon_p\}$ has a piecewise smooth boundary and $\bar{\partial}\phi = 0$ near $\Delta_\epsilon \cap \mathrm{supp}(\phi)$, we have the two following identities

$$< \bar{\partial}\frac{1}{f^{\beta+\underline{1}}}, [\phi] > = \frac{(p-1)!(-1)^{\frac{p(p-1)}{2}}}{(2\pi i)^p} \int_{\partial\Delta_\epsilon} \frac{\sum_{k=1}^p (-1)^{k-1} \bar{f}_k^{\beta_k+1} \overline{\partial f_{[k]}} \wedge \phi}{[\sum_{k=1}^p |f_j|^{2(\beta_j+1)}]^p} . \tag{3.32}$$

and

$$< \bar{\partial} \frac{1}{f^{\beta+\underline{1}}}, [\phi] >=$$

$$= \frac{(p-1)!(-1)^{\frac{p(p-1)}{2}}}{(2\pi i)^p} \frac{(p+|\beta|-1)!}{\beta!} \int_{\partial\Delta_\epsilon} \frac{\bar{f}^{*\beta} \sum_{k=1}^{p} (-1)^{k-1} \bar{f}_k \overline{df_{[k]}} \wedge \phi}{\|f\|^{2(p+|\beta|)}}.$$

(3.33)

Such formulas have already been obtained in Chapter 2, with df instead of $d\zeta$. The first one is a consequence of the fact that

$$\bar{\partial} \left(|f|^{*2\mu(\beta+\underline{1})} \frac{\sum_{k=1}^{p}(-1)^{k-1} \bar{f}_k^{\beta_k+1} \overline{\partial f_{[k]}^{\beta+\underline{1}}} \wedge \phi}{\left(\sum_{k=1}^{p} |f_j|^{2(\beta_j+1)}\right)^p} \right) = p\mu |f|^{*2\mu(\beta+\underline{1})} \frac{\overline{\partial f^{\beta+\underline{1}}} \wedge \phi}{\|f^{\beta+\underline{1}}\|^{2p}}.$$

The second one follows from

$$\bar{\partial} \left(|f_1 \cdots f_p|^{2\mu} \bar{f}^{*\beta} \frac{\sum_{k=1}^{p}(-1)^{k-1} \bar{f}_k \overline{\partial f_{[k]}} \wedge \phi}{\|f\|^{2(p+|\beta|)}} \right) = p\mu |f_1 \cdots f_p|^{2\mu} \bar{f}^{*\beta} \frac{\overline{\partial f} \wedge \phi}{\|f\|^{2(p+|\beta|)}}.$$

§4 Residue currents

At this point, the first question that could be asked is whether the residue of Grothendieck is a $(0,p)$ current supported by $\{f_1 = \ldots = f_p = 0\}$, in other words, can its action be extended to all $(n, n-p)$ test forms, not only $\bar{\partial}$-closed ones. We already know some candidates, the currents $R_t[f_1, \ldots, f_p]$ introduced in Proposition 3.6. In this direction, we can prove here the following result.

Theorem 3.18. Let $f_1, \ldots, f_p$ be p holomorphic functions defining a complete intersection in an open set $U \subset \mathbb{C}^n$. The currents $R_t[f_1, \ldots, f_p]$ are independent of $t \in (\mathbb{R}^+)^p$ and define a current annihilated by the ideal generated by $f_1, \ldots, f_p$ in $\Lambda_c^{(n,n-p)}(U)$; this current is called the residue current attached to $(f_1, \ldots, f_p)$; its action is usually denoted as

$$\phi \mapsto < \bar{\partial} \frac{1}{f_1} \wedge \cdots \wedge \bar{\partial} \frac{1}{f_p}, \phi >$$

Proof. The proof of this result lies on an argument of division of differential forms. The problem here is that this argument has to be carried out on the desingularization of the hypersurface $\{f_1 \cdots f_p = 0\}$ above a neighborhood of the origin (since the statement is of local nature, it will be enough for us to keep U as being some small neighborhood of the origin.). The parallel of this method with Leray's method we quoted before (and which works when the hypersurfaces $\{f_j = 0\}$ are in general position) will be easy to make here. It would be very interesting to obtain a proof of Theorem 3.18 independent of the resolution of singularities. We assume here that we have a desingularization map π for $f_1 \cdots f_p$ above the neighborhood in which we intend to prove our result; the notations we take here are exactly those used in the proof of Proposition 3.6.

We first prove that $R_1[f_1, \ldots, f_p]$ is annihilated by the ideal generated by $f_1, \ldots, f_p$ in $\mathcal{D}(U)$. More precisely, we prove that for any $\phi \in \Lambda_c^{(n, n-p)}(U)$, $R_1[f_1, \ldots, f_p](f_j \phi) = 0, j = 1 \ldots, p$. We will show that for $j = 1$. For that purpose, given a test form ϕ, let us consider the function F of two complex variables λ_1, λ_2 defined for Re $\lambda_1 \gg 0$, Re $\lambda_2 \gg 0$ by

$$
\begin{aligned}
F(\lambda_1, \lambda_2) &= \lambda_1 \lambda_2^{p-1} \int |f_1|^{2(\lambda_1 - 1)} |f_2 \ldots f_p|^{2(\lambda_2 - 1)} \overline{\partial f} \wedge \phi \\
&= (-1)^p \lambda_2^{p-1} \int \frac{|f_1|^{2\lambda_1}}{f_1} |f_2 \cdots f_p|^{2(\lambda_2 - 1)} \overline{\partial f_2} \wedge \cdots \wedge \overline{\partial f_p} \wedge \bar{\partial}(f_1 \phi) \\
&= (-1)^p \lambda_2^{p-1} \int |f_1|^{2\lambda_1} |f_2 \ldots f_p|^{2\lambda_2} \frac{1}{f_2 \cdots f_p} \frac{\overline{\partial f_2} \wedge \ldots \wedge \overline{\partial f_p}}{\bar{f}_2 \cdots \bar{f}_p} \wedge \bar{\partial}\phi.
\end{aligned}
$$
(3.34)

From now on, we will write:

$$
\bar{\partial}\psi = \sum_\tau \xi_\tau \wedge \bar{\omega}_\tau
$$

where ξ_τ are some $(n, 0)$ forms, ω_τ some $(n - p + 1, 0)$ forms with constant coefficients. We now place ourselves in a local chart for the desingularization manifold $\mathcal{X}$ (exactly as in the proof of proposition 3.6). Assuming that in this chart $\pi^* f_j = u_j w_1^{\alpha_{j,1}} \ldots w_n^{\alpha_{j,n}} = u_j w^{*\alpha_j}$, we express $F(\lambda_1, \lambda_2)$ as linear combination of terms of two types. The first, and most important, are of the form

$$
\lambda_2^{p-1} \int |w|^{*2(\lambda_1 \alpha_1 + \lambda_2 \beta)} \frac{1}{w^{*\beta}} \frac{d\bar{w}_{j_2}}{\bar{w}_{j_2}} \wedge \cdots \wedge \frac{d\bar{w}_{j_p}}{\bar{w}_{j_p}} \wedge \theta(w, \lambda) \overline{\pi^*(\omega_\tau)} \wedge \xi_\tau \quad (3.35)
$$

where $\mathcal{A} = \{j_2, \ldots, j_p\}$ is subset of $\{1, \ldots, n\}$ such that w_{j_2} appears in $\pi^* f_2, \ldots,$ w_{j_p} in $\pi^* f_p$. The function $(w, \lambda) \mapsto \theta(w, \lambda)$ is C^∞ in both variables, in $\mathcal{D}$ as a function of w, and entire as a function of λ; β is an element in $\mathbb{N}^n$ such that β_j is different from zero as soon as w_j is present in the expression of at least one of the functions $\pi^* f_2, \ldots, \pi^* f_p$.

The terms of the second type consist of those terms where the integrand contains at most $p-2$ distinct $\bar{w}_j$ in the denominator. The corresponding w_j appear in the monomial decomposition of the product $f_2 \cdots f_p$. A typical such term is

$$\lambda_2^{p-1} \int |w|^{*2(\lambda_1 \alpha_1 + \lambda_2 \beta)} \frac{1}{w^{*\beta}} \frac{1}{\bar{w}_{j_1} \cdots \bar{w}_{j_k}} \theta(w, \lambda) \, d\bar{w} \wedge dw \,, \qquad (3.36)$$

with $k \leq p-2$, the function θ has the same properties as the corresponding function in (3.35), and all $\beta_{j_l} \neq 0$ for $1 \leq l \leq k$.

We are going to study now one term of the form (3.35) and recall that we denoted as $\mathcal{A}$ the set $\{j_1, \ldots, j_p\}$ associated to such a term. One can write $\overline{\pi^* \omega_\tau}$ as

$$\overline{\pi^* \omega_\tau} = \sum_{\substack{\mathcal{A}' \subseteq \{1, \ldots, n\} \\ \#\mathcal{A}' = n-p+1}} \bar{\omega}_{\tau, \mathcal{A}'} \, d\bar{w}_{\mathcal{A}'} \,.$$

where the functions $\omega_{\tau, \mathcal{A}'}$ are holomorphic in the local chart and

$$d\bar{w}_{\mathcal{A}'} = \bigwedge_{l=1}^m d\bar{w}_{k_l}$$

if $m = \#\mathcal{A}'$ and $\mathcal{A}' = k_1, \ldots, j_m$ (in an increasing order.) We just use here the holomorphy of π and the fact that ω_τ has constant coefficients. The only non zero term which contributes to (3.35) is

$$\lambda_2^{p-1} \int |w|^{*2(\alpha_1 \lambda_1 + \beta \lambda_2)} \frac{1}{w^{*\beta}} \frac{d\bar{w}_{j_2}}{\bar{w}_{j_2}} \wedge \cdots \wedge \frac{d\bar{w}_{j_{i_p}}}{\bar{w}_{j_p}} \wedge \theta(w, \lambda) \bar{\omega}_{\tau, \mathcal{A}^c} \wedge d\bar{w}_{\mathcal{A}^c} \wedge \xi_\tau. \quad (3.37)$$

Here the notation $\mathcal{A}^c$ means $\{1, \ldots, n\} \setminus \mathcal{A}$. Let $P_{\mathcal{A}} = \{w : w_j = 0, j \in \mathcal{A}\}$. We distinguish two cases:

(i) We assume in this case that $\pi(\Gamma_{\mathcal{A}}) \subseteq \{f_1 = f_2 = \cdots = f_p = 0\}$. Since ω_τ is an $(n-p+1, 0)$ form with constant coefficients, it is automatically zero on the $n-p$ dimensional analytic set $\{f_1 = f_2 = \ldots = f_p = 0\}$ for reasons of dimensionality. Therefore, the restriction $\pi^* \omega_\tau|_{P_{\mathcal{A}}} = 0$, in other words, $\omega_{\tau, \mathcal{A}^c}$ vanishes on $\{w_{j_2} = \ldots = w_{j_p} = 0\}$. So that,

$$\omega_{\tau, \mathcal{A}^c} = w_{j_2} y_2 + \cdots + w_{j_p} y_p \,,$$

where $y_2, \ldots, y_p$ are holomorphic functions of w the chart. Thus, using integration by parts, as in the proof of Proposition 3.6, we represent (3.37) as a combination of terms of the form

$$\frac{\lambda_2^{p-1}}{\prod_s (\rho_s \lambda_1 + \sigma_s \lambda_2)} h(\lambda) \,, \qquad (3.38)$$

with h holomorphic near the origin in $\mathbb{C}^2$, the product in (3.38) has at most $p-2$ factors, and every factor satisfies $\sigma_s \neq 0$ (in fact, $\sigma_s \in \{\beta_{j_2},\ldots,\beta_{j_p}\}$.)

(ii) In this case, $\pi(P_A) \not\subset \{f_1 = f_2 = \ldots = f_p = 0\}$. Since all functions $\pi^*(f_j)$, $2 \leq j \leq n$, vanish on P_A, f_1 cannot vanish on $\pi(P_A)$. This implies that $w_{j_2},\ldots,w_{j_p}$ do not appear in the monomial decomposition of $\pi^* f_1$. Using again integration by parts, to decrease the singularities, one can now represent (3.37) as a combination of terms of the form

$$\frac{\lambda_2^{p-1}}{\prod\limits_s (\sigma_s \lambda_2)} k(\lambda), \tag{3.39}$$

where k is holomorphic at the origin in $\mathbb{C}^2$ and the product in (3.39) consists of at most $p-1$ factors, each of them with $\sigma_s \neq 0$ (here again, $\sigma_s \in \{\beta_{j_2},\ldots,\beta_{j_p}\}$.)

Let us consider now a term of the form (3.36). Since the number of variables $\bar{w}_j$ in the denominator of the integrand is strictly smaller than $p-1$, and since every one of these variables appears in at least one of the monomials $\pi^*(f_2),\ldots,\pi^*(f_p)$, we have exactly the same situation as in the case (i) considered earlier. Thus, we can express this term as a linear combination of terms of the form (3.38).

Summarizing, it is clear that, for Re $\lambda_1 \gg 0$, Re $\lambda_2 \gg 0$, $F(\lambda)$ can be written as

$$F_1(\lambda) + \lambda_2^{p-1} \sum_s \frac{h_s(\lambda)}{g_s(\lambda)}, \tag{3.40}$$

where the sum is finite, F_1, and the h_s are holomorphic near the origin in $\mathbb{C}^2$, and each g_s is a product of at most $p-2$ linear factors of the form $\rho\lambda_1 + \sigma\lambda_2$, with $\sigma \neq 0$. Clearly, $\lambda_2 \mapsto F(0,\lambda_2)$ makes sense for Re $\lambda_2 \gg 0$ (this can be seen from formula (3.34)). Moreover, this function is the integral over $\mathbb{C}^n$ of a $\bar{\partial}$-closed form with compact support. So, for Re $\lambda_2 \gg 0$, $F(0,\lambda_2) \equiv 0$. The fact that F can be expressed in the form (3.40) shows that it makes sense to consider $\lambda_2 \mapsto F(0,\lambda_2)$ for $\lambda_2 \neq 0$; putting together the fact that $F(0,\lambda_2) = F_1(0,\lambda_2) + \lambda_2 F_2(\lambda_2)$, where F_2 is holomorphic at the origin, and the vanishing of $\lambda_2 \mapsto F(0,\lambda_2)$ for Re $\lambda_2 \gg 0$, we see that $F_1(\underline{0}) = F(\underline{0}) = 0$ (as a consequence of the analytic continuation.) This proves that $R_{\underline{1}}$ is annihilated by f_1, hence by all the f_j.

The fact that $R_t[f_1,\ldots,f_p]$ is independent of t can be proved via the same argument. It is enough show that for any $t_1,t_1',t_2,\ldots,t_p$ in $\mathbb{R}^+$, we have, for $\phi \in \Lambda_c^{n,n-p}(U)$,

$$R_{t_1,t_2,\ldots,t_p}[f_1,\ldots,f_p](\phi) = R_{t_1',t_2,\ldots,t_p}[f_1,\ldots,f_p](\phi). \tag{3.41}$$

Note that these two values can be obtained from the restrictions to two different half-lines in the first orthant of a meromorphic function G of two variables. G

is defined, for $\operatorname{Re} \lambda_1 \gg 0$, $\operatorname{Re} \lambda_2 \gg 0$, by

$$G(\lambda) = \frac{(-1)^{\frac{p(p-1)}{2}}}{(2\pi i)^p} \lambda_1 \lambda_2^{p-1} \int |f_1|^{2(\lambda_1 - 1)} |f_2 \cdots f_p|^{2(\lambda_2 - 1)} \overline{\partial f} \wedge \phi,$$

which equals, using Stokes's theorem, to

$$\frac{(-1)^{\frac{p(p-1)}{2} + p}}{(2\pi i)^p} \lambda_2^{p-1} \int \frac{|f_1|^{2\lambda_1}}{f_1} |f_2 \cdots f_p|^{2(\lambda_2 - 1)} \overline{\partial f_2} \wedge \cdots \wedge \overline{\partial f_p} \wedge \bar{\partial}\phi.$$

Exactly the same reasoning that led to the form (3.40) for F, shows that the function G can be written as

$$G(\lambda) = G_1(\lambda) + \lambda_2^{p-1} \sum_s \frac{h_s(\lambda)}{g_s(\lambda)},$$

here g_s, h_s have the same properties as in (3.40). In fact, the only difference between F and G lies in the term $1/f_1$ in the integrand, but all what we did above was to eliminate antiholomorphic singular factors. It is clear hence that the values at $\mu = 0$ of $G(t_1\mu, t_2\mu)$ and $G(t_1'\mu, t_2\mu)$ coincide. This implies the equality (3.41) and concludes the proof of Theorem 3.18. □

Remark 3.19. The proof of this theorem was taken from [3, Theorem 1.3]. It is a transposition to the context of analytic continuation of distributions of an argument originally developed by Coleff and Herrera [11], and more recently by Passare [32]. Coleff and Herrera proposed to define the residue current as

$$\lim_{\delta \to 0} \left(\frac{1}{(2\pi i)^p} \int_{\Gamma_f(\epsilon(\delta))} \frac{\phi}{f_1 \cdots f_p} \right),$$

where $\delta \mapsto (\epsilon_1(\delta), \ldots, \epsilon_p(\delta))$ avoids critical values and tends to zero when δ tends to zero in a somehow restricted way (admissible paths). Namely, for any $m > 0$ and any $1 \leq j \leq p - 1$,

$$\epsilon_j(\delta) = O(\epsilon_{j+1}(\delta)^m).$$

Passare introduced another way to multiply principal values and residue currents [33]. He considered the average of the limit values of integrals over the cycles $\Gamma_f(\delta^{s_1}, \ldots, \delta^{s_p})$. This method allows the residue current to be approximated by smooth forms.

For the division problems we are interested in, the method of analytic continuation of the powers, because of its relation to the theory of $\mathcal{D}$-modules, appears to be more useful.

§5 The local duality theorem

When $f_1, \ldots, f_p$, define a complete intersection in an open set $U \subset \mathbb{C}^n$, the Grothendieck residue can be used as a test for the membership problem relative to the ideal generated by $f_1, \ldots, f_p$ in $H(U)$. Namely, we have the following theorem, originally due to R. Harvey [17], which can be considered as a local duality theorem. We give here its C^∞ version.

Definition 3.20. Let I be an ideal in the algebra $H(U)$, U open subset of $\mathbb{C}^n$, then the *local ideal* I_{loc} is the family of all functions $f \in H(U)$ such that for each $\zeta_0 \in U$ there are a neighborhood N_{ζ_0} of ζ_0, a finite collection of functions $g_j \in I$ $(j \in J)$, and another collection $\phi_j \in H(N_{\zeta_0})$ so that

$$f(\zeta) = \sum_{j \in J} g_j(\zeta)\phi_j(\zeta) \quad \forall \zeta \in N_{\zeta_0}. \tag{3.42}$$

This condition says that for every point ζ_0 in U, the germ f_{ζ_0} belongs to the ideal I_{ζ_0} generated by I in the local ring $_n\mathcal{O}_{\zeta_0}$. If U is pseudoconvex, then $I_{loc} = I$ (see [21].) One can also replace $_n\mathcal{O}_{\zeta_0}$ by the ring $C_{\zeta_0}^\infty$ of germs of smooth functions at ζ_0 and obtain an ideal $\mathcal{I}_{loc} \subseteq C^\infty(U)$. A flatness argument shows that $\mathcal{I}_{loc} \cap H(U) = I_{loc}$ (see [37].)

For a finite family $f_1, \ldots, f_m$ in $H(U)$, we denote by $I(f_1, \ldots, f_m)$ the ideal they generate in $H(U)$.

Theorem 3.21. *Let $f_1, \ldots, f_p$, be p holomorphic functions defining a complete intersection in an open set $U \subset \mathbb{C}^n$, and let $I = I(f_1, \ldots, f_p)$. For $h \in H(U)$ the following two properties are equivalent:*

(i) $h \in I_{loc}$.

(ii) $\forall \phi \in \Lambda_c^{n,n-p}(U)$, $< \bar{\partial}\frac{1}{f_1} \wedge \ldots \wedge \bar{\partial}\frac{1}{f_p}, h\phi >= 0$.

Remark 3.22. Later on, we give a cohomological version of this theorem, namely, we show that (i) and (ii) are also equivalent to

(iii) $\forall \phi \in \Lambda_c^{n,n-p}(U)$ if $\bar{\partial}\phi = 0$ near $\{f_1 = \ldots = f_p = 0\}$ then $< \bar{\partial}\frac{1}{f}, h\phi >= 0$

Let us point out that the equivalence between (i) and (iii) is stronger than Theorem 3.21, but the interest of the following proof lies in the division formulas we develop. Everything in the proof below is done semilocally, hence it can be extended to any strictly pseudoconvex with a C^1 boundary.

Proof of Theorem 3.21. We have already proved that condition (i) implies (ii) in Theorem 3.18. To prove the converse, since it is a local property, we may assume that U is a ball $B(0, r)$, such that $f_1, \ldots, f_p$ are holomorphic in a bigger ball $U' = B(0, r')$, $r' > r$, and they define in U' a system in *normal position*, that is, for any $j_1 < \ldots < j_k$ in $\{1, \ldots, p\}$ and any $\zeta_0 \in \{\zeta \in U' : f_{j_1} = \ldots = f_{j_k} = 0\}$, we have $\dim_{\zeta_0}\{f_{j_1} = \ldots = f_{j_k} = 0\} = n - k$ (see [16, p. 660].)

There exist functions $g_{j,k}$, $j = 1,\ldots,p$, $k = 1,\ldots,n$, holomorphic in $U' \times U'$, so that, for every $1 \le j \le p$ and every $z, \zeta \in U'$, one has

$$f_j(z) - f_j(\zeta) = \sum_{k=1}^{n} g_{j,k}(z,\zeta)(z_k - \zeta_k) = < \vec{g_j}(z,\zeta), z - \zeta >,$$

where $\vec{g_j}$ is the vector of coordinates $g_{j,k}$. The $g_{j,k}$ can be written down explicitly using Taylor's formula with integral remainder term. (For U' pseudoconvex, the existence of the $g_{j,k}$ can be found in [21, Theorem 7.2.9]. This formula is usually called Hefer's formula.)

We are going to use now the Cauchy formula with two weights, as given in Theorem 2.7. We keep the notation from the proof of this theorem. Let $\mu \in \{\text{Re } \mu \gg 0\}$ and $\epsilon > 0$. Consider the two $\mathbb{C}^n$-valued maps in $C^1(\bar{U} \times \bar{U})$,

$$q_1(z,\zeta) := q_1(z,\zeta;\mu) := \frac{1}{p} \sum_{j=1}^{p} |f_j(\zeta)|^{2(\mu-1)} \bar{f}_j(\zeta) \vec{g_j}(z,\zeta)$$

$$q_2(z,\zeta) := q_2(\zeta;\epsilon) := \frac{1}{\rho(\zeta) - \epsilon} \left(\frac{\partial \rho}{\partial \zeta_1}, \ldots, \frac{\partial \rho}{\partial \zeta_n} \right)(\zeta) = \frac{\vec{\partial}\rho(\zeta)}{\rho(\zeta) - \epsilon}$$

where $\rho(\zeta) = \|\zeta\|^2 - r^2$. Recall that we frequently leave the variable ζ implicit. Let $G_1(t) = P(t)$, with P a polynomial of degree p, satisfying $P(1) = 1$, to be chosen later, and $G_2(t) = t^{-N}$, where N is a positive integer, which will also be made explicit below. Let

$$\Phi_{1,\mu}(z,\zeta) := 1+ < q_1(z,\zeta;\mu), z - \zeta >$$
$$= \frac{1}{p} \sum_{j=1}^{p} (1 - |f_j|^{2\mu}) + \frac{1}{p} \sum_{j=1}^{p} |f_j|^{2(\mu-1)} \bar{f}_j f_j(z). \qquad (3.43)$$

$$\Phi_{2,\epsilon}(z,\zeta) := 1+ < q_2(\zeta;\epsilon), z - \zeta >= \frac{\rho - \epsilon+ < \vec{\partial}\rho, z - \zeta >}{\rho - \epsilon}. \qquad (3.44)$$

Note that given $z \in U$, there exists an open neighborhood $W \subset\subset U$ of z such that G_2 is holomorphic on a neighborhood of the image of $\bar{W} \times \bar{U}$ by the maps $\Phi_{2,\epsilon}$. From Theorem 2.7, for $z \in W$ we have

$$h(z) = \frac{1}{(2\pi i)^n} \left(\int_U h(\zeta) P_{\epsilon,\mu}(z,\zeta) + \int_{\partial U} h(\zeta) K_{\epsilon,\mu}(z,\zeta) \right), \qquad (3.45)$$

where, we recall, the kernels K, P can be written as

$$K_{\epsilon,\mu}(z,\zeta) = \sum_{\alpha_0 + |\alpha| = n-1} \frac{1}{\alpha!} \Gamma_{1,\mu}^{(\alpha_1)} \Gamma_{2,\epsilon}^{(\alpha_2)} \frac{S \wedge (\bar{\partial}S)^{\alpha_0} \wedge \bar{\partial}Q_{1,\mu}^{\alpha_1} \wedge \bar{\partial}Q_{2,\epsilon}^{\alpha_2}}{\|z - \zeta\|^{2(\alpha_0+1)}}$$

$$P_{\epsilon,\mu}(z,\zeta) = \sum_{|\alpha|=n} \frac{1}{\alpha!} \Gamma_{1,\mu}^{(\alpha_1)} \Gamma_{2,\epsilon}^{(\alpha_2)} \bar{\partial}Q_{1,\mu}^{\alpha_1} \wedge \bar{\partial}Q_{2,\epsilon}^{\alpha_2},$$

with

$$S(z, \zeta) = \sum_{k=1}^{n} (\bar{\zeta}_k - \bar{z}_k) d\zeta_k,$$

$$Q_{1,\mu}(z, \zeta) = \sum_{k=1}^{n} q_{1,k}(z, \zeta; \mu) d\zeta_k,$$

$$Q_{2,\epsilon}(z, \zeta) = \frac{1}{\rho(\zeta) - \epsilon} \sum_{k=1}^{n} \frac{\partial \rho}{\partial \zeta_k}(\zeta) d\zeta_k = \frac{\partial \rho}{\rho(\zeta) - \epsilon},$$

and the functions $\Gamma_{1,\mu}^{(\alpha_1)}$, $\Gamma_{2,\epsilon}^{(\alpha_2)}$ are defined starting with the two pairs $(q_{1,\mu}, G_1)$, $(q_{2,\epsilon}, G_2)$ as it was done in Theorem 2.7.

Due to our choice of G_2, we see that when $N \geq n$, $\rho(\zeta) - \epsilon$ appears only with nonnegative exponents in the kernel $P_{\epsilon,\mu}$ and strictly positive exponents in $K_{\epsilon,\mu}$. Thus, it makes sense to let $\epsilon = 0$ in formula (3.45), provided that Re $\mu \gg 0$. As the kernel $K_{0,\mu}$ vanishes on ∂U, we are left with the representation formula

$$h(z) = \frac{1}{(2\pi i)^n} \int_U h(\zeta) P_\mu(z, \zeta), \qquad (3.46)$$

where $P_\mu = P_{0,\mu}$ can be written as

$$P_\mu(z, \zeta) = \sum_{k=0}^{p} P^{(k)}(\Phi_1(z, \zeta; \mu)) \, B_k \wedge \bar{\partial} Q_{1,\mu}^k, \qquad (3.47)$$

with an $(n-k, n-k)$ form B_k, smooth on U, given by

$$B_k = B_k(z, \zeta) =$$

$$= (-1)^{n-k} \frac{(N+n-k-1)!}{k!(n-k)!(N-1)!} \left(\frac{\rho}{<\vec{\partial}\rho, z - \zeta> + \rho} \right)^{N+n-k} (\bar{\partial}\partial \log \rho)^{n-k}.$$

If $N > n+1$ and we define the coefficients of B_k as being zero outside U, they become functions of class $\mathcal{C}^{N-n-1}$ in $\mathbb{C}^n$.

The idea we are going to use now is that, given a finite family $\mathcal{F}$ of real analytic non negative functions in U' then, for any $c \in \mathbb{R}$, there is a non-negative integer L, which depends on c, U, and $\mathcal{F}$, such that for $F \in \mathcal{F}$, the analytic continuation of $\mu \mapsto F^\mu$ from Re $\mu > 0$ into Re $\mu > c$, is a meromorphic function with values in $\mathcal{D}_L'(\bar{U})$, the space of distributions of order at most L in $\bar{U}$ (see [2].) We use this observation with $c = -n - 1$ and the family $\mathcal{F}$ consists of the single function $|f_1 \cdots f_p|^2$. We choose $N > L + n + 1$. This choice allows us to analytically continue the right-hand side of (3.46) to the half-plane $\{$Re $\lambda > -1\}$. Hence, $h(z)$ can be identified to the coefficient of μ^0

in the Laurent development (of the analytic continuation) of $\int_U h(\zeta)P_\mu(z,\zeta)$ at $\mu = 0$.

Some of the terms involved in $\int_U h(\zeta)P_\mu(z,\zeta)$ give a contribution to $h(z)$ that belong to the ideal I, others do not. Let as point out that because P is a polynomial of degree p, the term corresponding to $k = p$ in (3.47) can be written as $\kappa B_p \wedge \bar\partial Q^p_{1,\mu}$, with κ a positive constant depending on P. On the other hand, if we introduce the $(1,0)$ differential forms

$$g_j := g_j(z,\zeta) := \sum_{k=1}^n g_{j,k}(z,\zeta)d\zeta_k, \ 1 \le j \le p,$$

then

$$\bar\partial Q^p_{1,\mu} = (-1)^{\frac{p(p-1)}{2}} \frac{p!}{p^p} \mu^p |f_1 \dots f_p|^{2(\mu-1)} \overline{\partial f} \wedge g_1 \wedge \dots \wedge g_p.$$

The contribution of $\kappa \int_U h B_p \wedge \bar\partial Q^p_{1,\mu}$ to (3.46) equals

$$\kappa' < \bar\partial \frac{1}{f}, \ hg_1 \wedge \cdots \wedge g_p \wedge B_p > . \tag{3.48}$$

From Theorem 3.18 we conclude that the expression (3.48) vanishes, since the residue current is annihilated by $h \in I_{loc}$.

We have still to study in detail those other terms, which, a priori, do not give a contribution that belongs to the ideal I_{loc}. These are terms of the form

$$\mu^k \int_U hP^{(k)} \left(\frac{1}{p} \sum_{j=1}^p (1 - |f_j|^{2\mu})\right) |f_{j_1}|^{2(\mu-1)} \dots |f_{j_k}|^{2(\mu-1)} \overline{\partial f_{j_1}} \wedge \dots \wedge \overline{\partial f_{j_k}} \wedge \omega,$$

$$\tag{3.49}$$

where $1 \le k \le p - 1$, $J := \{j_1 < \dots < j_k\}$, $1 \le j_l \le p$, and ω is a compactly supported $(n, n - k)$ form of class C^L. In order to study (3.49) we use the binomial formula to obtain a linear combination of terms of the type

$$\int_U h \prod_{j \in J'} (1 - |f_j|^{2\mu}) \wedge \bar\partial \left[\frac{|f_j|^{2\mu}}{f_j}\right] \wedge \omega, \tag{3.50}$$

where $J' \subset \{1, \dots, p\}$, $\#J' \le p - k$. Note that each term of the wedge product in (3.50) contributes one power of μ, this shows that the factor μ^k in (3.49) has been absorbed into this wedge product. We claim that, if $J' \not\subset J$, then the contribution of (3.50) is zero. In fact, let $j'_0 \in J' \setminus J$, and consider, as in the proof of Theorem 3.18, the function of $\lambda \in \mathbb{C}^2$

$$F(\lambda) = \int_U h(1 - |f_{j_0}|^{2\lambda_1}) \left(\prod_{j' \in J' \setminus \{j'_0\}} (1 - |f_{j'}|^{2\lambda_2})\right) \wedge \bar\partial \left(\frac{|f_j|^{2\lambda_2}}{f_j}\right) \wedge \omega.$$

Because $f_1, \ldots, f_p$ are in normal position in the ball U', the functions $f_l, l \in J \cup J'$ (there may be repetitions but we do not take them into account) define also a complete intersection in U'. The argument we used in Theorem 3.18 to prove that the residue was annihilated by the ideal, shows that F can be written in the form (3.40), with k instead of p, and so that $F(0,0) = 0$ (as $F(0, \lambda_2) \equiv 0$ for Re $\lambda_2 \gg 0$.)

Thus, we can replace (3.49) by

$$\mu^k \int_U h P^{(k)} \left(\frac{1}{p} \sum_{j \in J} (1 - |f_j|^{2\mu}) \right) \bigwedge_{j \in J} |f_j|^{2(\mu-1)} \overline{\partial f_j} \wedge \omega. \tag{3.51}$$

To fix the ideas let us take $J = \{1, \ldots, k\}$. For any positive integer m, we have

$$\mu^k \int_U h (1 - |f_1|^{2\mu})^m |f_1 \cdots f_k|^{2(\mu-1)} \bigwedge_{j \in J} \overline{\partial f_j} \wedge \omega =$$

$$= \mu^k \int_U \left(\sum_{l=0}^m \binom{m}{l} (-1)^l |f_1|^{2((l+1)\mu-1)} \right) h |f_2 \cdots f_k|^{2(\mu-1)} \bigwedge_{j \in J} \overline{\partial f_j} \wedge \omega. \tag{3.52}$$

From the independence of $R_t[f_1, \ldots, f_k](h\omega)$ from t, we see that the contribution of (3.52) is

$$(-1)^{\frac{k(k-1)}{2}} (2\pi i)^k \sum_{l=0}^m \binom{m}{l} (-1)^l \frac{1}{l+1} R_{\underline{1}}[f_1, \ldots, f_k](h\omega).$$

More generally, the contribution of (3.51) is, up to the factor $(-1)^{\frac{k(k-1)}{2}} (2\pi i)^k$,

$$\left(\int_0^1 \cdots \int_0^1 P^{(k)} \left(\frac{1}{p} \left(\sum_{j=1}^k x_j \right) \right) dm_k(x) \right) < \bar{\partial} \frac{1}{f_{j_1}} \wedge \ldots \wedge \bar{\partial} \frac{1}{f_{j_k}}, h\omega >$$

Integration by parts shows that

$$\int_0^1 \cdots \int_0^1 P^{(k)} \left(\frac{1}{p} \left(\sum_{j=1}^k x_j \right) \right) dm_k(x) = p^k \sum_{j=1}^k (-1)^{k-j} \binom{k}{j} P\left(\frac{j}{p} \right). \tag{3.53}$$

The particular choice of polynomial P given by

$$P(t) = \frac{1}{p!} \prod_{j=0}^{p-1} (pt - j),$$

satisfies $P(1) = 1$ and, for any k, $1 \le k \le p - 1$,

$$\sum_{j=1}^{k}(-1)^{k-j}\binom{k}{j}P(\frac{j}{p}) = 0 .$$

This clever choice of P implies that the only terms in (3.46) that contribute to the value $h(z)$ are those that belong to the ideal I. This concludes the proof that $h \in I$.   □

Remark 3.23. Observe that the last proof, when applied to an arbitrary function $h \in H(U')$, where U' is a neighborhood of $\bar{U}$, U bounded strictly pseudoconvex domain, provides a division formula with remainder (cf., Chapter 1, §4.) Namely, for any $z \in U$,

$$h(z) = \sum_{j=1}^{p} h_j(z)f_j(z) + \frac{p!}{(2\pi i)^{n-p}} < \bar{\partial}\frac{1}{f}, \bigwedge_{j=1}^{p} g_j \wedge B_p > . \tag{3.54}$$

Moreover, one can see that the quotient h_{j_0}, is given by contributions at $\mu = 0$ of linear combinations of expressions of one of the following form; either of the form

$$\left(\int_{U} h\frac{|f|^{*2\mu(\alpha+\beta)}}{f^{*\beta}} \bigwedge_{j\in J}\left((\bar{\partial}\frac{|f_j|^{2\mu}}{f_j}) \wedge g_j\right) \wedge B_k\right)\left(\prod_{j\neq j_0} f_j(z)^{\beta_j}\right) f_{j_0}^{\beta_{j_0}-1}(z),$$

where $\#J = k$, $\alpha, \beta \in \mathbb{N}^p$, $\beta_{j_0} > 0$, $|\alpha| + |\beta| \le p - k$; either of the form

$$\left(\int_{U} h\frac{|f|^{*2\mu(\alpha+\beta)}}{f^{*\beta}} B_0\right)\left(\prod_{j\neq j_0} f_j(z)^{\beta_j}\right) f_{j_0}^{\beta_{j_0}-1}(z),$$

where again $\alpha, \beta \in \mathbb{N}^p$, $\beta_{j_0} > 0$, $|\alpha|+|\beta| \le p-k$; remark that these expressions can be written in the form

$$< hT, \bigwedge_{j\in J} g_j \wedge B_k > (\prod_{j\neq j_0} f_j^{\beta_j}(z)) f_{j_0}^{\beta_{j_0}-1}(z) ,$$

or

$$< hT, B_0 > (\prod_{j\neq j_0} f_j^{\beta_j}(z)) f_{j_0}^{\beta_{j_0}-1}(z) ,$$

where $\bar{\partial}T = f^{*a}R$, R being a Grothendieck residue associated to $k+1$ monomial expressions in $f_1, \ldots, f_p$, which still define a complete intersection (which is possible since $k < p$.)

The fact that $f_1, \ldots, f_p$ define a complete intersection in U played a crucial role in the proof of the last result. Nevertheless, similar ideas can be used to prove the Briançon-Skoda theorem [10]. Before proceeding, we need to recall a few definitions from commutative algebra.

Definition 3.24. *Let I be an ideal in a commutative ring $\mathcal{R}$. An element $h \in \mathcal{R}$ ia algebraically dependent on I if it satisfies a monic relation*

$$h^k + a_1 h^{k-1} + \ldots + a_k = 0, \ a_l \in I^l, \ 1 \le l \le k.$$

Example. Let $\mathcal{R} =_n \mathcal{O}_0$ and $I = I(f_1, \ldots, f_p)$, where $f_1, \ldots, f_p$ belong to the maximal ideal of the local ring $_n\mathcal{O}_0$. If h is algebraically dependent on I, there exists a neighborhood ω of 0 and a constant $C > 0$ such that, if we choose representatives for $f_1, \ldots, f_p, h$ in $H(\omega)$, we have

$$|h(\zeta)| \le C \max_{1 \le j \le p} |f_j(\zeta)|, \quad \forall \zeta \in \omega. \tag{3.55}$$

Consider the graded subring $\mathcal{I} = \sum_{l \ge 0} I^l T^l$ of $\mathcal{R}[T]$. The fact that h is algebraically dependent on I is equivalent to the fact that hT satisfies a monic equation $P(hT) = 0$, with the coefficients of P in $\mathcal{I}$. This implies that the set of algebraic elements over an ideal I is also an ideal in $\mathcal{R}$, denoted $\bar{I}$ [31].

Integral dependence can be checked, when $\mathcal{R}$ is an integral domain, using the valuative criterion of integral dependence ([41], vol II, p. 353-354). Let us recall this criterion. A discrete valuation on $\mathcal{R}$ is a map $\nu : \mathcal{R} \mapsto \mathbb{N} \cup \{\infty\}$ such that, if $\mathcal{R}^*$ denotes the family of invertible elements of $\mathcal{R}$, then

$$\nu(xy) = \nu(x) + \nu(y), \quad \nu(x+y) \ge \min(\nu(x), \nu(y)),$$
$$\nu(\mathcal{R}^*) = 0, \quad \nu(0) = \infty$$

An interesting example of valuation in the ring $_n\mathcal{O}_0$ is to take as $\nu(f)$ the order of vanishing of f at 0, i.e., the degree of the homogeneous polynomial $p \not\equiv 0$ such that

$$f(\zeta) = p(\zeta) + o(\|\zeta\|^{\deg(p)})$$

When $\mathcal{R}$ is an integral domain and I an ideal in $\mathcal{R}$, then $h \in \mathcal{R}$ belongs to $\bar{I}$ if and only if, for any discrete valuation ν on $\mathcal{R}$, we have

$$\nu(h) \ge \min\{\nu(a) : a \in I\}.$$

In the context of germs of analytic functions, this amounts to say that $h \in \bar{I}(f_1, \ldots, f_p)$ if and only if, for any germ of a curve $\gamma \in (_1\mathcal{O}_0)^n$, the germ $h \circ \gamma$ is in the ideal generated by $f_1 \circ \gamma, \ldots, f_p \circ \gamma$ in $\mathbb{C}\{t\}$ (the ring of formal power series) (see [24].) As an application of this criterion, one can see that if $f \in {}_n\mathcal{O}_0$, $f(0) = 0$, and $df(z) = 0$ implies that $f(z) = 0$, that is, 0 is an isolated critical point of the function f, then f is algebraically dependent on the Jacobian ideal $I(\partial f/\partial z_1, \ldots, \partial f/\partial z_n)$. In 1974, Briançon and Skoda proved the following theorem about algebraic dependence.

Theorem 3.25. *Let* $I = I(f_1, \ldots, f_p)$ *be an ideal in* $_n\mathcal{O}_0$. *Then, for* $s = \min(p, n)$, *we have* $\bar{I}^s \subseteq I$.

Proof. The original proof of this theorem [10] was based on L^2-estimations. We give here a different proof, in some sense more effective, that is, it provides a division formula, as in Chapter 1, §4, and also it leads to estimates for the constant C in (3.55) in terms of the behaviour of some principal value distributions.

The idea is to write a division formula (with remainder) for any element $h \in H(U)$, $U = B(0, r)$, a ball about the origin, such that $h, f_1, \ldots, f_p$ are holomorphic in a neighborhood U' of $\bar{U}$. We keep the same notation as in the previous proof, and introduce the Hefer vectors $\vec{g_j} = (g_{j,1}, \ldots, g_{j,n})$ so that for all $z, \zeta \in U'$

$$f_j(z) - f_j(\zeta) = \sum_{k=1}^{n} g_{j,k}(z, \zeta)(z_k - \zeta_k).$$

Let $\mu \in \{\text{Re } \mu \gg 0\}$, $\epsilon \in \mathbb{R}^+$, and two C^1 maps from $\bar{U} \times \bar{U}$ into $\mathbb{C}^n$,

$$\tilde{q}_1(z, \zeta) = \tilde{q}_1(z, \zeta; \mu) := \|f\|^{2(\mu-1)} (\sum_{j=1}^{p} \bar{f}_j(\zeta) \vec{g_j}(z, \zeta)$$

$$q_2(z, \zeta) = q_2(z, \zeta; \epsilon) := \frac{1}{\rho(\zeta) - \epsilon} \vec{\partial}\rho(\zeta) \text{ where } \rho(\zeta) = \|\zeta\|^2 - r^2.$$

Let $\sigma = \min(p, n+1)$, $G_1(t) = t^\sigma$, and $G_2(t) = t^{-N}$ (N to be chosen later.) The reasoning followed during the proof of Theorem 3.23 can be carried over in this case also. The only difference lies in the fact we now need to deal with the analytic continuation of

$$\mu \mapsto \int \|f\|^{2\mu} \phi,$$

where ϕ is a test form. The existence of the analytic continuation of this map, as a distribution-valued meromorphic function, as well as the remark on the order L of these distributions stated during the proof of the previous theorem, still holds here. We need presently to be a little more precise, in order to keep track of the complex analytic nature of the problem. Apply Hironaka's resolution of singularities for the complex hypersurface $\{f_1 \cdots f_p = 0\}$, so that, in a local chart, $\pi^*(f_j) = u_j w^{*\alpha_j}$. From what we have already seen, the problem is to study in this local chart of the desingularization $\mathcal{X}$, the analytic continuation of the integral

$$\int (|u_1 w^{*\alpha_1}|^2 + \ldots + |u_1 w^{*\alpha_p}|^2)^\mu \psi(w) d\bar{w} \wedge dw. \tag{3.56}$$

Here ψ is a test function in $\mathcal{D}(\mathbb{C}^n)$, $u_1, \ldots, u_p$ are nonvanishing holomorphic functions in a neighborhood of $\mathrm{supp}(\psi)$, and $\alpha_1, \ldots, \alpha_p$ are elements in $\mathbb{N}^n$.

Let us introduce the by now classical ideas of Varchenko [40] and Khovanskii [23], and introduce the Newton polyhedron $\Gamma^+(\alpha_1, \ldots, \alpha_p)$ in $(\overline{\mathbb{R}^+})^n$, defined as the convex hull of the set

$$\bigcup_j \{\alpha_j + (\overline{\mathbb{R}^+})^n\}.$$

One associates to this polyhedron an algebraic n-dimensional manifold $\mathcal{X}(\Gamma^+)$, as follows: consider the set of the M normal directions to the $n-1$-dimensional faces of this Newton polyhedron, and represent it by a system $\mathcal{S}$ of vectors $\rho_1, \ldots, \rho_M$, with integral non-negative coordinates. Then, using an algorithm due to Mumford, et al. [30], one completes the system $\mathcal{S}$ into a minimal system $\tilde{\mathcal{S}}$ of vectors with non-negative integral coordinates, which provides a partition of the first orthant into a finite number of distinct cones (with edges along the elements $\tilde{\rho}_j$ of $\tilde{\mathcal{S}}$; furthermore, this system satisfies the following property: if $\tilde{\rho}_{j_1}, \ldots, \tilde{\rho}_{j_n}$ are the edges of one of the cones of this partition, then

$$\det(\tilde{\rho}_{j_1}, \ldots, \tilde{\rho}_{j_n}) = \pm 1.$$

A standard terminology for such a partition is to call it a *fan*.

One can construct an algebraic variety $\tilde{\mathcal{X}}(\Gamma^+)$ (a *toroidal variety*) by gluing together different copies $\mathcal{U}_J$ (indexed by the set of cones of the fan) of $\mathbb{C}^n$, parametrized by the monoidal transformations:

$$\tilde{\pi}_J : \mathcal{U}_J \to \mathbb{C}^n, \ \tilde{\pi}_J(t_1, \ldots, t_n) = (t_1^{\tilde{\rho}_{j_1,1}} \ldots t_n^{\tilde{\rho}_{j_n,1}}, \ldots, t_1^{\tilde{\rho}_{j_1,n}} \ldots t_n^{\tilde{\rho}_{j_n,n}}).$$

Two points (one in $\mathcal{U}_J$, the other in $\mathcal{U}_{J'}$) are equivalent if and only if the monoidal map from $\mathcal{U}_J$ to $\mathcal{U}_{J'}$ given by $\tilde{\pi}_{J'}^{-1} \circ \tilde{\pi}_J$ is defined at the first point and takes this point to the second one. Glueing the copies $\mathcal{U}_J$ together induces a map $\tilde{\pi}$ in $\mathcal{X}(\Gamma^+)$. The map $\tilde{\pi}$ is a proper map from $\tilde{\mathcal{X}}(\Gamma^+)$ into $\mathbb{C}^n$. Moreover, it is invertible from $\tilde{\mathcal{X}}(\Gamma^+) \setminus \tilde{\pi}^{-1}(\{w_1 \cdots w_n = 0\})$ to $\mathbb{C}^n \setminus \{w_1 \cdots w_n = 0\}$. Furthermore, and this is for us the main point, if we express $w^{*\alpha_j}$, $1 \leq j \leq p$, in a local chart $\mathcal{U}$ in $\tilde{\mathcal{X}}(\Gamma^+)$, then the monomials $\tilde{\pi}^*(w^{*\alpha_j})$ are monomials in the new coordinates t, and all these monomials are multiples of a distinguished one among them; we will denote this distinguished monomial by m.

We are now ready to conclude the proof of Theorem 3.25. As in the proof of Theorem 3.21, we choose the exponent N in the definition of G_2 sufficiently large, namely $N > L' + n + 1$, where L' is a non-negative integer such the analytic continuation of $\mu \mapsto \|f\|^{2\mu}$ from $\mathrm{Re}\,\mu > 0$ to $\mathrm{Re}\,\mu > -\sigma - 1$ exists in $\mathcal{D}'_{L'}(\bar{U})$; then, for $h \in H(U')$ and $z \in U$,

$$h(z) = \frac{1}{(2\pi i)^n} \int_U h(\zeta) \tilde{P}_\mu(z, \zeta), \qquad (3.57)$$

where P_μ can be written as

$$\tilde{P}_\mu(z,\zeta) = \sum_{k=0}^{s} \frac{\sigma!}{(\sigma-k)!} \left(1 - \|f\|^{2\mu} + \|f\|^{2(\mu-1)} \sum_{j=1}^{p} \bar{f}_j f_j(z) \right)^{\sigma-k} B_k \wedge \bar{\partial} \tilde{Q}_{1,\mu}^k.$$

with

$$\tilde{Q}_{1,\mu}(z,\zeta) = \sum_{j=1}^{p} \tilde{q}_{1,k}(z,\zeta;\mu)d\zeta_k$$

and the differential forms B_k are the same as in the proof of the previous theorem. An easy computation leads to

$$\bar{\partial}\tilde{Q}_{1,\mu}^k = \|f\|^{2(\mu-1)k}\theta^k + k(\mu-1)\|f\|^{2((\mu-1)k-1)}\bar{\partial}\|f\|^2 \wedge \sum_{j=1}^{p} \bar{f}_j g_j \wedge \theta^{k-1},$$

where the $(1,0)$ differential forms g_j are associated to Hefer systems of divisors as before, and the $(1,1)$ differential form θ is given by

$$\theta(z,\zeta) = \sum_{j=1}^{p} \overline{\partial f_j} \wedge g_j(z,\zeta).$$

For $k = p$ the last computation can be considerably simplified. We set $\psi_{j,\mu} = \psi_j := \|f\|^{2(\mu-1)}\bar{f}_j$. Therefore,

$$\tilde{Q}_{1,\mu} = \sum_{j=1}^{p} \psi_j g_j$$

and, since the g_j are $\bar{\partial}$-closed,

$$\partial Q_{1,\mu}^p = (-1)^{\frac{p(p-1)}{2}} p! \bigwedge_{j=1}^{p} \bar{\partial}\psi_j \wedge \bigwedge_{j=1}^{p} g_j .$$

We also have,

$$\bar{\partial}\psi_j = (\mu-1)\bar{f}_j\|f\|^{2(\mu-2)}\bar{\partial}\|f\|^2 + \|f\|^{2(\mu-1)}\overline{\partial f_j}$$

Hence,

$$\bigwedge_{j=1}^{p} \bar{\partial}\psi_j = \|f\|^{2p(\mu-1)}\overline{\partial f} + (\mu-1)\|f\|^{2(p-1)(\mu-1)}\|f\|^{2(\mu-2)}\|f\|^2\overline{\partial f}$$

$$= \mu\|f\|^{2p(\mu-1)}\overline{\partial f} .$$

Thus,

$$\bar{\partial}\tilde{Q}^p_{1,\mu} = (-1)^{\frac{p(p-1)}{2}} p!\mu\|f\|^{2p(\mu-1)}\overline{\partial f} \wedge \bigwedge_{j=1}^{p} g_j .$$

We now consider the analytic continuation of both sides in (3.57) and we identify $h(z)$ with the Laurent coefficient of μ^0 in the expansion of $\int_U h(\zeta)\tilde{P}_\mu(z,\zeta)$. It is immediate to notice that the only terms giving some contribution which does not lie, as functions of z, in the ideal generated by $f_1,\ldots,f_p$ are of the form

$$\mu\frac{c(k,s)}{(2\pi i)^n} \int_U h(1-\|f\|^{2\mu})^{\sigma-k}\|f\|^{2((\mu-1)k-1)}\bar{\partial}\|f\|^2\wedge(\sum_{j=1}^{p}\bar{f}_j g_j)\wedge\theta^{k-1}\wedge B_k \quad (3.58)$$

for some k such that $1 \le k \le s$. One has to explain this in detail; first note that any expression of one of the following forms

$$\int_U h(1-\|f\|^{2\mu})^{\sigma-k}\|f\|^{2(\mu-1)k}\theta^k \wedge B_k,$$

$$\int_U h(1-\|f\|^{2\mu})^{\sigma-k}\|f\|^{2((\mu-1)k-1)}\bar{\partial}\|f\|^2 \wedge (\sum_{j=1}^{p}\bar{f}_j g_j) \wedge \theta^{k-1} \wedge B_k,$$

for some k such that $1 \le k < \sigma$,

$$\int_U h(1-\|f\|^{2\mu})^{\sigma} \wedge B_0,$$

does not contribute to the Laurent coefficient of μ^0 at the origin; the only remaining expression to consider occurs in the case when $p \le n$ and comes from the term

$$s! \int_U hB_s\bar{\partial}\tilde{Q}^s_{1,\mu},$$

which is of the form (3.58) because of the explicit form of $\bar{\partial}\tilde{Q}^s_{1,\mu}$. Let us now study a term of the form (3.58). As in the proof of Proposition 3.6, we first transform it using Hironaka's theorem. We know that the desingularization map is proper, so that the set $\pi^{-1}(\text{supp}(\chi_U B_k))$ is compact; one can consider a covering of it with local charts of $\mathcal{X}$ (and a partition of unity associated with this covering.) We now reduce (via pull-back) our problem to one of these charts, which we suppose parametrized by a system of coordinates w, centered at $w = 0$, so that $\pi^* f_1,\ldots,\pi^* f_p$ are of the form $u_j w^{*\alpha_j}$; let us denote as $w \mapsto \xi(w)$ the fonction from the partition of unity with support in the corresponding chart. We now consider the algebraic manifold $\tilde{\mathcal{X}}(\Gamma^+_{\tilde{\alpha}})$ relative to $\alpha_1, \ldots, \alpha_p$ and its associate proper map $\tilde{\pi}_\alpha$. Consider the pull-back of the differential form

$$\mu h \|f\|^{2((\mu-1)k-1)} \bar{\partial} \|f\|^2 \wedge \sum_{j=1}^{p} \bar{f}_j g_j \wedge \theta^{k-1} \wedge \chi_U B_k$$

on this chart and multiply it by ξ. If one takes the pull back of this new form to an open set $\mathcal{U}_J$ relative to $\tilde{\mathcal{X}}(\Gamma_{\tilde{\alpha}}^+)$, we obtain a differential form of the type

$$\mu \tilde{\pi}_J^*(\xi^* h) \varphi_J |m|^{2k\mu} \frac{1}{m^k} \left(\overline{\frac{\partial m}{m}} \wedge \eta_1 + \eta_2 \right), \qquad (3.59)$$

where m is the distinguished monomial, η_1 and η_2 are differential forms with bounded coefficients (note that no regularity condition is required here for η_1 and η_2) and φ_J is a function with compact support relative to the partition of unity subordinated to the covering of $\tilde{\pi}_{\tilde{\alpha}}^{-1}(\text{supp}(\xi))$ by the different copies $\mathcal{U}_J$ of $\mathbb{C}^n$ which appear in the construction of $\mathcal{X}(\Gamma_{\tilde{\alpha}}^+)$. Let us now assume that we have a function $\tilde{h}$ which is in $H(U')$ and in the integral closure of $I(f_1, \ldots, f_p)$; then, there is a positive constant C such that, for any ζ in U, one has

$$|\tilde{h}(\zeta)| \le C \|f(\zeta)\|.$$

This implies, in the local chart on $\mathcal{X}$ where we are working, an inequality of the form

$$|\tilde{\pi}_J^*(\xi^* \tilde{h}^s) \varphi_J| \le C_{\alpha,J} |m|^s$$

for some positive constant $C_{\tilde{\alpha},J}$. Fom this estimate, one can see immediately that, when we take $h = \tilde{h}^s$ in our representation formula, any term of the form (3.58), which can be expressed as a combination of expressions such as

$$\mu \int ((\tilde{\pi}_J^*(\xi_\alpha \pi^* h))^s |m|^{2k'\mu} \frac{1}{m^k} [\overline{\frac{\partial m}{m}} \wedge \tilde{\eta}_1^{(k)} + \tilde{\eta}_1^{(k)}], \qquad (3.60)$$

with $k' \in \mathbb{N}$ and $\tilde{\eta}_1^{(k)}, \tilde{\eta}_2^{(k)}$ some compactly supported bounded differential forms, does not contribute to the Laurent coefficient of μ^0 at the origin in the right hand side of (3.57). This shows that our formula (3.57) provides an explicit representation of $\tilde{h}^s$ in the ideal generated by $f_1, \ldots, f_p$ in U. This concludes the proof of Theorem 3.25. $\qquad \square$

Remark 3.26. As before, what we did above provides an explicit division formula with a remainder term in a strictly pseudoconvex domain U. When h belongs to the integral closure of the ideal, our division formula can be written as

$$h^s(z) = \sum_{j=1}^{p} h_j(z) f_j(z), \qquad (3.61)$$

A careful look at the division formulas show that the quotients are somehow a bit simpler when one divides h^σ instead of h^s; the quotient h_{j_0} appears as a linear combination of Laurent coefficients of μ^0 (in the development around the origin) of expressions which are of the two following forms; either

$$\left(\int_U h^\sigma \|f\|^{2((\mu-1)(k+|\beta|)-1)} \Theta_k \wedge B_k \right) \left(\prod_{j \neq j_0} f_j(z)^{\beta_j} \right) f_{j_0}^{\beta_{j_0}-1}(z) \qquad (3.62)$$

for some $k \geq 1$ and

$$\Theta_k = \theta^{k-1} \wedge \left(\theta \|f\|^2 - k\bar{\partial}\|f\|^2 \wedge \sum_{j=1}^p \bar{f}_j g_j \right)$$

and $\beta \in \mathbb{N}^p$ such as $\beta_{j_0} > 0$ and $|\beta| + k \leq \sigma$, or of the form

$$\left(\int_U h^\sigma \|f\|^{2(\mu-1)|\beta|} B_0 \right) \left(\prod_{j \neq j_0} f_j(z)^{\beta_j} \right) f_{j_0}^{\beta_{j_0}-1}(z). \qquad (3.62')$$

for some $\beta \in \mathbb{N}^p$ such as $|\beta| \leq \sigma$ and $\beta_{j_0} > 0$.

Remark 3.27. The choice of $\tilde{q}_1(..;\lambda)$ is not the only one one can take in order to write down a division formula as in Theorem 3.25. For example, one can choose some element t in $(\mathbb{R}^+)^p$ and consider

$$q(\zeta, z; \mu, t) := \frac{|f(\zeta)|^{*2\mu t}}{\|f\|^2} \sum_1^p \overline{f_j(\zeta)} g_j(\zeta, z). \qquad (3.63)$$

If one uses such a weight, the division formula one obtains instead of (3.61) involves, in the representation of the quotients, the action of the distributions appearing in the polar part of the development at $\mu = 0$ of

$$\mu \longmapsto \frac{|f|^{*2\mu(t-\underline{1})}}{\|f\|^{2s}}$$

considered as a meromorphic distribution-valued map. We will see later on how the action of such distributions can be described in terms of Bernstein-Sato functional identities for germs of holomorphic functions.

When $f_1, \ldots, f_p$ define a complete intersection in a neighborhood of the origin (that is when $f_1, \ldots, f_p$ define a regular sequence in $_n\mathcal{O}_0$), we will see in the following chapter a much simpler proof of Briançon-Skoda Theorem based precisely on the local duality result. This proof we will propose there will be better adapted to the result itself, which is in fact a result of an algebraic nature.

It is not clear in the analytic proof we have just given that the Grothendieck residue is closely related to the Briançon-Skoda theorem; this was remarked by Lipman and Teissier in [26] and provided the key idea for an algebraic proof of this theorem (and therefore a generalization to a more abstract algebraic situation of this result.) What Lipman and Teissier proved is that the Briançon-Skoda's theorem is valid in any regular n dimensional local ring $\mathcal{R}$ (regular means that if $\bar{x}_1, \ldots, \bar{x}_n$ is a basis for $\mathcal{M}/\mathcal{M}^2$, where $\mathcal{M}$ is the maximal ideal, then the correspondence $\mathcal{K}[t_1, \ldots, t_n] \longrightarrow \mathrm{grad}(\mathcal{R})$, $t_i \mapsto \bar{x}_i$, with $\mathcal{K} = \mathcal{R}/\mathcal{M}$, is injective). As we will see in the next chapter, this result will be very useful for us in order to get effective bounds for some division questions in commutative algebra.

To complete this chapter, we would like to mention that the currents, which appear in the division formulas we wrote in Theorem 3.21 and Theorem 3.25, have the curious property of being annihilated by multiplication by the conjugates $\bar{f}_j$ of the generators . For the case of Theorem 3.21, this property is a consequence of the following result.

Proposition 3.27. Let $f_1, \ldots, f_p$ in $H(U)$ define a complete intersection, then for any $j \in \{1, \ldots, p\}$ and any $(n, n-p)$ test form ϕ,

$$< \bar{\partial} \frac{1}{f}, \bar{f}_j \phi >= 0 .$$

Proof. Let Re $\mu \gg 0$, and remark that

$$\mu^p \int_U |f_1 \cdots f_p|^{2(\mu-1)} \overline{\partial f} \wedge \bar{f}_j \phi = \pm \mu \left(\int_U \frac{|f_j|^{2\mu}}{f_j} \bigwedge_{k \neq j} \bar{\partial} \left[\frac{|f_k|^{2\mu}}{f_k} \right] \wedge \overline{\partial f_j} \wedge \phi \right). \quad (3.64)$$

It follows from Proposition 3.6 that the value at $\mu = 0$ of the analytic continuation of (3.64) is zero. $\qquad \square$

In the division formula (3.57) used in the proof of the Briançon-Skoda theorem, all the currents that appear in the remainder term have the form, after pull-back to a toroidal manifold,

$$\phi \mapsto \left(\mu \int |m|^{2k\mu} \frac{1}{m^k} \left(\frac{\overline{\partial m}}{\overline{m}} \wedge \theta_1 + \theta_2 \right) \wedge \phi \right)_{|\mu=0}, \quad (3.65)$$

where m is a monomial, $k \in \mathbb{N}$, θ_1, θ_2 are smooth forms, and the value at $\mu = 0$ is taken after analytic continuation. If in (3.65) we take $\phi = \bar{m}\psi$, its value is zero as a consequence of Proposition 3.6.

An important consequence of the fact that these currents are annihilated by the conjugates of the generators is that they are supported by the analytic variety

$$\{f_1 = \cdots = f_p = 0\},$$

instead of the hypersurface $\{f_1 \cdots f_p = 0\}$. In particular, in the case of complete intersection, the residue current is annihilated both by the generators and their conjugates.

References for Chapter 3

[1] L.A. Aizenberg and A.P. Yuzhakov, *Integral Representation in Multidimensional Complex Analysis*, Transl. Amer. Math. Soc. 58, 1980.

[2] M.F. Atiyah, Resolution of singularities and division of distributions, Commun. Pure Appl. Math. 23(1970), 145–150.

[3] C.A. Berenstein, R. Gay, and A. Yger, Analytic continuation of currents and division problems, Forum Math.1(1989), 15–51.

[4] C.A. Berenstein and A. Yger, Une formule de Jacobi et ses conséquences, Ann. Sci. Ec. Norm. Sup. Paris 24(1991), 363–377.

[5] C.A. Berenstein and A. Yger, Formules de représentation intégrale et problè- mes de division, in *Diophantine Approximations and Transcendental Numbers, Luminy 1990*, P. Philippon (ed.), Walter de Gruyter, Berlin 1992, 15–37.

[6] B. Berndtsson, Integral formulas on projective space and the Radon transform of Gindikin-Henkin-Polyakov, Publications Mathematiques 32(1988), 7–41.

[7] E. Bierstone and P.D. Milman, Uniformization of analytic spaces, J. Amer. Math. Soc. 2(1989), 801–836.

[8] J.-E. Björk, *Rings of Differential Operators*, North-Holland, Amsterdam 1979.

[9] J.-E. Björk, Regular holonomic distributions, Chapter 7 of unpublished manuscript.

[10] J. Briançon and H. Skoda, Sur la clôture intégrale d'un ideal de germes de fonctions holomorphes en un point de $\mathbb{C}^n$, Comptes Rendus Acad. Sci. Paris, sér. A Math. 278(1974), 949–951.

[11] N. Coleff and M. Herrera, *Les courants résidus associés à une forme méromorphe*, Lect. Notes in Math. 633, Springer Verlag, New York, 1978.

[12] P. Dolbeault, *Theory of residues and homology*, Lect. Notes in Math. 116, Springer-Verlag, New York, 1970.

[13] P. Dolbeault, General theory of multidimensional residues, in *Encyclopedia of Math. Sci.*, vol. 7 (*Several Complex Variables I*), Springer Verlag, 1990.

[14] H. Federer, *Geometric Measure Theory*, Springer-Verlag, New York, 1969.

[15] I.S. Gradshtein and I.M. Ryzhik, *Table of integrals, series, and products*, Academic Press, New York, 1980.

[16] P. Griffiths and J. Harris, *Principles of Algebraic Geometry*, Wiley-Interscience, New York, 1978.

[17] R. Harvey, Holomorphic chains and their boundaries, Proc. Symp. Pure Math. XXX, Amer. Math. Soc.,1977.

[18] M. Herrera and D. Lieberman, Residues and principal values on complex spaces, Math. Ann. 194(1971), 259–94.

[19] H. Hironaka, Resolution of singularities of an algebraic variety over a field of characteristic 0, I,II, Annals of Math. 79(1964), 109–326.

[20] L. Hörmander, On the division of distributions by polynomials, Arkiv Math. 3(1958), 555–568.

[21] L. Hörmander, *An Introduction to Complex Analysis in Several Complex Variables*, North Holland, Amsterdam, 1973.

[22] M. Kashiwara and T. Kawai, On holonomic systems for $(f_1 + \sqrt{-1}\,0)^{\lambda_1} \cdots (f_N + \sqrt{-1}\,0)^{\lambda_N}$, Publ. Res. Inst. Math. Sci., Kyoto Univ., 15(1979), 551–575.

[23] A.G. Khovanskii, Newton polyhedra and toroidal varieties, Funct. Anal. Appl., 11(1978), 289–295.

[24] M. Lejeune and B. Teissier, Clôture intégrale des idéaux et équisingularité, Publ. de l'Inst. Fourier, St. Martin d'Hères, 1975.

[25] J. Lipman, *Residues and Traces of Differential Forms via Hochshild Homology*, Contemporary Math. 61, Amer. Math. Soc., 1987.

[26] J. Lipman and B. Teissier, Pseudorational local rings and a theorem of Briançon-Skoda about integral closures of ideals, Michigan Math. J. 28(1981), 97–115.

[27] S. Lojasiewicz, Sur le problème de la division, Studia Math. 18(1959), 87–136.

[28] J. Leray, Le calcul différentiel et intégral sur une variété analytique complexe, problème de Cauchy III, Bull. Soc. Math. France 87(1959), 81–180.

[29] B. Lichtin, Generalized Dirichlet series and B-functions, Compositio Math. 65(1988), 81–120.

[30] D. Mumford, G. Kempf, B. St. Donnat, and F. Knudsen, Toroidal embedding, Lect. Notes in Math. 339, Springer-Verlag, New York, 1973.

[31] D.G. Northcott and D. Rees, Reduction of ideals in local rings, Proc. Cambr. Phil. Soc. 50(1954), 145–158.

[32] M. Passare, Residues, currents, and their relation to ideals of holomorphic functions, Math. Scand. 62(1988), 75–152.

[33] M. Passare, Courants méromorphes et égalité de la valeur principale et de la partie finie, Lect. Notes in Math. 1295, Springer-Verlag, New York, 1987, 157–166.

[34] M. Passare and M. Andersson, A shortcut to weighted representation formulas for holomorphic functions, Ark. Math. 26(1988), 1–12.

[35] C. Sabbah, Proximité évanescente I, Compositio Math. 62(1987), 283–328; C. Sabbah, Proximité évanescente II. Equations fonctionnelles pour plusieurs fonctions analytiques, Compositio Math. 64(1987), 213–241.

[36] L. Schwartz, Division par une fonction holomorphe sur une variété analytique complexe, Summa Brasil. Math. 3(1955), 181–209.

[37] J.C. Tougeron, *Idéaux de fonctions différentiables*, Springer-Verlag, New York, 1972.

[38] A.K. Tsikh, *Multidimensional Residues and Their Applications*, Transl. Amer. Math. Soc. 103, 1992.

[39] B.L. van der Warden, *Modern Algebra*, Springer-Verlag, New York, 1979.

[40] A.N. Varchenko, Newton polyhedra and estimation of oscillating integrals, Funct. Anal. Appl. 10(1976), 175–196.

[41] O. Zariski and P. Samuel, *Commutative Algebra*, Springer-Verlag, New York, 1958.

Chapter 4
The Cauchy-Weil Formula and its Consequences

§1 The Cauchy-Weil formula

In this chapter we shall give several applications of the well-known Cauchy-Weil formula [17], [18], and we shall point out its relation to the Grothendieck residue introduced in Chapter 3. We start this section by showing that the Cauchy-Weil formula, at least in some cases, can be derived from the Cauchy formula with weights, proved in Theorem 2.7.

Let U be an open domain in $\mathbb{C}^n$ and consider functions $f_1, \ldots, f_p$ holomorphic in U, $p \geq n$. Let U_j, $1 \leq j \leq p$, be relatively compact domains in the complex plane, with the property that $\bar{U}_j \subseteq f_j(U)$. The set

$$D := \{\zeta \in U : f_j(\zeta) \in U_j, j = 1, \ldots, p\}$$

is called an analytic polyhedron.

Definition 4.1. *A relatively compact connected component Δ of the set D in U, satisfying the additional properties that all the boundaries ∂U_j are piecewise smooth and, for any $k \in \{1, \ldots, p\}$, the intersections of k of the faces*

$$\gamma_j := \{\zeta \in \bar{\Delta}, f_j(\zeta) \in \partial U_j, f_l(\zeta) \in U_l \text{ for } l \neq j\}$$

are either empty, or piecewise smooth submanifolds of dimension $2n - k$ in $\mathbb{R}^{2n}$, is called a Weil polyhedron.

A situation where these conditions hold, occurs when

$$\text{rank} \begin{pmatrix} \text{grad}(f_{j_1}) \\ \ldots \\ \text{grad}(f_{j_k}) \end{pmatrix} = k \quad \text{on} \quad \gamma_{j_1} \cap \ldots \cap \gamma_{j_k},$$

for any $k \leq n$ and any $j_1 < \ldots < j_k$. For example, if $f_1, \ldots, f_n$ define a complete intersection in U, we have seen in Chapter 3, §3, that, for almost all choices of $\epsilon_1, \ldots, \epsilon_n > 0$, and for $U_j = D(0, \epsilon_j)$, any relatively compact connected component Δ_ϵ of $D (= D(\epsilon))$ in U is a Weil polyhedron.

The boundary of a Weil polyhedron Δ consists of a finite number of $(2n - 1)$-dimensional hypersurfaces $\sigma_j \subseteq \gamma_j$. The orientation of Δ, as an open subset of $\mathbb{C}^n$, induces an orientation on the faces σ_j. The orientation of each σ_j induces an orientation for its $(2n - 2)$-dimensional faces $\sigma_{j,l} \subseteq \gamma_{j,l} := \gamma_j \cap \gamma_l$, $j \neq l$, in such a way that the identity between cycles $\partial \sigma_j = \sum \sigma_{j,l}$ holds. Continuing this procedure may lead to some empty faces, but in the remaining cases one can iterate this process n times and therefore define n-dimensional oriented surfaces

$\sigma_{j_1,\dots,j_n}$. The cycle $\sum \sigma_{j_1,\dots,j_n}$ (with its orientation) is called the skeleton of the Weil polyhedron. We denote below by J, an arbitrary ordered multiindex $j_1 < \dots < j_n$, and by σ_J, the corresponding n-dimensional face. We remark that a Weil polyhedron Δ is always a pseudoconvex domain and, moreover, the Hefer functions mentioned in Chapter 3, §5, are defined in a neighborhood of $\bar{\Delta} \times \bar{\Delta}$.

The Cauchy-Weil representation formula is given in following theorem.

Theorem 4.2. *Let Δ be a Weil polyhedron with respect to $U, U_1, \dots, U_p$, and $f_1, \dots, f_p$. Let $\bar{g}_j$ (respectively, g_j) be the corresponding vectors of Hefer functions (resp., differential forms.) For any $z \in \Delta$, for any $f \in H(\Delta) \cap C(\bar{\Delta})$,*

$$f(z) = \frac{(-1)^{\frac{n(n-1)}{2}}}{(2\pi i)^n} \sum_{j \in J} \int_{\sigma_J} \frac{f(\zeta) \bigwedge_{j \in J} g_j(z, \zeta)}{(f_{j_1}(\zeta) - f_{j_1}(z)) \cdots (f_{j_n}(\zeta) - f_{j_n}(z))} \qquad (4.1)$$

The proof is very standard and it can be found, e.g., in [1], [17]. In the particular case when $p = n$ and $\Delta = \Delta_\epsilon$ is a Weil polyhedron associated to a system of functions $f_1, \dots, f_n$, defining a complete intersection in U, the Cauchy-Weil formula allows us to expand any function $h \in H(\Delta) \cap C(\bar{\Delta})$, as a power series in $f_1, \dots, f_n$. We have, for $z \in \Delta$,

$$h(z) = \frac{(-1)^{\frac{n(n-1)}{2}}}{(2\pi i)^n} \sum_{k \in \mathbb{N}^n} \left(\int_{\Gamma_f(\epsilon)} \frac{h(\zeta) \bigwedge_{j=1}^n g_j(z, \zeta)}{f(\zeta)^{*(k+\underline{1})}} \right) f(z)^{*k}. \qquad (4.2)$$

The series in (4.2) is uniformly and absolutely convergent on any compact subset of Δ (we have used the notation $\Gamma_f(\epsilon)$ for the cycle $\{|f_1| = \epsilon_1, \dots, |f_n| = \epsilon_n\}$, which was defined in (3.18).) The proof of formula (4.2) from (4.1) is exactly the same as the proof that the Cauchy formula implies any holomorphic function can be represented by a power series. Note that the coefficient of $f(z)^{*k}$ in (4.2) equals

$$< \bar{\partial} \frac{1}{f(\zeta)^{*(k+\underline{1})}}, h(\zeta)\theta(\zeta) \bigwedge_{j=1}^n g_j(z, \zeta) >,$$

where θ is any test function which equals 1 on a neighborhood of Δ. This follows from the results of Chapter 3, §3. Let us give a similar statement for bounded domains U instead of Weil polyhedra.

Proposition 4.3. *Let U be a bounded domain with a piecewise smooth boundary in $\mathbb{C}^n$, and $f_1, \dots, f_n$ be holomorphic functions in a pseudoconvex neighborhood of $\bar{U}$, without common zeros on ∂U; then the set $V := \{\zeta \in U : f_1(\zeta) = \cdots = f_n(\zeta) = 0\}$ is finite and, for any function $h \in H(U) \cap C(\bar{U})$ and any test* *function* θ, *which is identically equal to* 1 *in a neighborhood of* V, *we have the following representation formula*

$$h(z) = \lim_{L \to \infty} \sum_{k \in \mathbb{N}^n, |k| \le L} < \bar{\partial} \frac{1}{f^{*(k+\underline{1})}}(\zeta), h(\zeta)\theta(\zeta) \bigwedge_{1 \le j \le n} g_j(z,\zeta) > f(z)^{*k},$$

(4.3)

valid for those $z \in U$ *such that*

$$\|f(z)\| < \min\{\|f(\zeta)\| : \zeta \in \partial U\}.$$

(4.4)

The convergence is uniform on any compact subset of U *where* (4.4) *holds.* (*Here the* g_j *are the differential forms associated to a system of Hefer functions in a neighborhood of* $\bar{U}$.)

Proof. The proof depends on the weighted formulas of Theorem 2.7. We keep the notation from that theorem. Here we use the case of two pairs of weights; the first one is given as (q_μ, G) and is defined as follows. For Re $\mu \gg 0$,

$$q_\mu(z,\zeta) = \frac{|f(\zeta)|^{*\mu\underline{2}}}{\|f\|^2} \sum_{j=1}^{n} \overline{f_j(\zeta)}\vec{g}_j(z,\zeta)$$

and $G(t) = t^N$, for an integer $N \ge n$. In order to define the second pair of weights, we introduce some function χ in $\mathcal{D}(U)$ which is identically equal to one in some neighborhood of z. The n-valued map $q = q^{(\chi)}$ is defined as

$$q^{(\chi)}(\zeta, z) = (1 - \chi(\zeta))\frac{\bar{\zeta} - \bar{z}}{\|\zeta - z\|^2}$$

and the associated map G_0 is $t \mapsto G_0(t) = t^{n+1}$. For any $z \in U$ we have

$$h(z) = \frac{1}{(2\pi i)^n}\left(\int_U h(\zeta)P_\mu^{(1)}(z,\zeta;N) + \int_U h(\zeta)P_\mu^{(2)}(z,\zeta;N)\right)$$

(4.5)

where

$$P_\mu^{(1)}(z,\zeta;N) = \binom{N}{n}\chi(\zeta)^{n+1}\left[1 - |f(\zeta)|^{*\mu\underline{2}} + |f(\zeta)|^{*\mu\underline{2}}\frac{\sum_{j=1}^{n}\overline{f_j(\zeta)}f_j(z)}{\|f(\zeta)\|^2}\right]^{N-n}\bar{\partial}Q_\mu^n$$

(4.6)

and

$$P_\mu^{(2)}(z, \zeta; N) =$$

$$= \sum_{\substack{\alpha_0+\alpha_1=n \\ \alpha_0>0}} \binom{N}{\alpha_1}\binom{n+1}{\alpha_0}\left[1 - |f(\zeta)|^{*\mu\underline{2}} + |f(\zeta)|^{*\mu\underline{2}}\frac{\sum_{j=1}^{n}\overline{f_j(\zeta)}f_j(z)}{\|f(\zeta)\|^2}\right]^{N-\alpha_1} \times$$

$$\times (\chi(\zeta))^{n+1-\alpha_0}(\bar{\partial}Q^{(\chi)})^{\alpha_0} \wedge \bar{\partial}Q_\mu^{\alpha_1}.$$

(4.7)

The kernels we are using are similar to those in the proofs of Theorem 2.7 (the weight $q^{(\chi)}$) and of Theorem 3.25 (the weight q_μ.) The differential forms Q^χ and Q_μ are associated to these weights as in the preliminaries of Chapter 3, §3.

Let us first study the analytic continuation of $P_\mu^{(1)}$ to $\mu = 0$. The same computation as that of Theorem 3.25 for the n-th power of $\bar\partial \tilde{Q}_{1,\mu}$ gives

$$\bar\partial Q_\mu^n = (-1)^{\frac{n(n-1)}{2}} n! n \mu |f(\zeta)|^{*n\mu 2} \frac{\overline{\partial f}}{\|f\|^{2n}} \wedge \bigwedge_{j=1}^n g_j . \tag{4.8}$$

Replace (4.8) into (4.6), and use the binomial formula to expand the bracketed expression in (4.6), to obtain

$$\sum_{l=0}^{N-n} \binom{N-n}{l} \frac{|f|^{*l\mu 2}}{\|f\|^{2l}} \left(\sum_{r=0}^{N-n-l} \binom{N-n-l}{r} (-1)^r |f|^{*r\mu 2} \right) \left(\sum_{|k|=l} \frac{l!}{k!} \bar{f}^{*k} f(z)^{*k} \right). \tag{4.9}$$

Fix l, fix a multiindex k with $|k| = l$, and fix an index r in (4.9). The corresponding term in (4.5) (for $P_\mu^{(1)}$) is holomorphic at $\mu = 0$, and its value can be computed, in terms of residue currents, using Proposition 3.16. This value equals

$$c_{l,k,r} < \bar\partial \frac{1}{f^{*(k+\underline{1})}}, h\chi^{n+1} \bigwedge_{j=1}^n g_j >, \tag{4.10}$$

where

$$c_{l,k,r} = (-1)^j \frac{N!}{(N-n)!} \frac{l!}{k!} \binom{N-n}{l} \binom{N-n-l}{r} \left(\frac{n+l}{n+l+r} \right) \left(\frac{k!}{(n+l)!} \right).$$

A simple combinatorial argument shows that for l, k fixed,

$$\sum_{r=0}^{N-n-l} c_{l,k,r} = 1.$$

We can replace in (4.10) the function χ^{n+1} by any any test function in U, which is identically 1 near V, the value of the residue does not change. Thus, the contribution of the first integral in (4.5) to the value $h(z)$ equals

$$\sum_{|k|\leq N-n} < \bar\partial \frac{1}{f^{*(k+\underline{1})}}, h\theta \bigwedge_{j=1}^n g_j > f(z)^{*k} . \tag{4.11}$$

Let us now study the second integral in (4.5). First, rewrite the form Q_μ as $|f|^{*\mu 2} A$, for some differential form A, singular only on the variety V. Hence, for any $0 \leq \alpha_1 \leq n - 1$,

$$\bar\partial Q_\mu^{\alpha_1} = |f|^{*\mu \alpha_1 2} \bar\partial A^{\alpha_1} + \mu \alpha_1 \frac{|f|^{*\mu 2}}{\overline{f^{*\underline{1}}}} \overline{\partial f^{*\underline{1}}} \wedge A \wedge \bar\partial A^{\alpha_1-1}. \tag{4.12}$$

We substitute this expression into (4.7) and we compute the analytic continuation to $\mu = 0$. Remark that in (2.13) we have already computed the powers of $\bar{\partial}Q^{(\chi)}$. If we substitute them into (4.7), we see that every term of the kernel $P_\mu^{(2)}$ contains as a factor either $(1-\chi)^\beta\chi^\gamma$, with $\beta, \gamma > 0$, or $(1-\chi)^\beta\bar{\partial}\chi$, with $\beta > 0$. Therefore, A has no singularities on the domain of integration. Thus, the contribution of the term with μ as a factor vanishes at $\mu = 0$, since the singularity arises only from antiholomorphic denominators and one can apply Proposition 3.6 to resolve it. For the remaining term, we can just let $\mu = 0$, since there is no singularity. Hence, as in (2.15), we can let $\chi \to \chi_U$ and the contribution of the second term becomes

$$\frac{1}{(2\pi i)^n}\int_{\partial U} h \sum_{\alpha_0+\alpha_1=n-1}\binom{N}{\alpha_1}\left(\frac{\sum_{j=1}^n \overline{f_j}f_j(z)}{\|f\|^2}\right)^{N-\alpha_1}\frac{S\wedge\bar{\partial}S^{\alpha_0}\wedge\bar{\partial}A^{\alpha_1}}{\|z-\zeta\|^{2(\alpha_0+1)}} \quad (4.13)$$

where, as usual, $S(z,\zeta) = \sum_{k=1}^n(\bar{\zeta}_k - \bar{z}_k)d\zeta_k$. The term between parentheses is dominated by

$$\frac{\|f(z)\|}{\min\{\|f(\zeta)\| : \zeta \in \partial U\}} < 1$$

by the hypothesis (4.4) (and clearly uniformly away from 1 over compact subsets of the open subset of U defined by (4.4).) Moreover, the combinatorial coefficient is dominated by $N^n/n!$, so that when $N \to \infty$ the term (4.13) goes to zero. This concludes the proof of Proposition 4.3. $\square$

It will be useful to have a version of this proposition with weights. Keeping the same notation as above, we additionally introduce a function τ with values in $\mathbb{C}^n$, of class $\mathcal{C}^1(\bar{U}\times\bar{U})$, which will play a role similar to that of θ in the operators R_j^θ of Chapter 1, §4. As in Proposition 2.8, for any $h \in H(U)$, we consider the function

$$\Gamma_h(z,\zeta) = \Gamma_h(z,\zeta;\tau) = h(\zeta)+ <\tau(\zeta), \zeta - z> .$$

We also introduce the differential forms

$$T(z,\zeta) = \sum_{j=1}^n \tau_j(z,\zeta)d\zeta_j ,$$

$$g(z,\zeta) = \bigwedge_{j=1}^n g_j(z,\zeta), \quad g_{[j]}(z,\zeta) = \bigwedge_{\substack{j=1\\j\neq k}}^n g_j(z,\zeta),$$

$$S(z,\zeta) = \sum_{k=1}^n(\bar{\zeta}_k - \bar{z}_k)d\zeta_k ,$$

$$\Theta(z,\zeta) = (\Gamma_h g)(z,\zeta) + (-1)^{n-1}\left(\sum_{j=1}^n(-1)^j(f_j(z) - f_j(\zeta))g_{[j]}(z,\zeta)\right)\wedge T(z,\zeta).$$

Finally, we recall from the last proof, that

$$A(z,\varsigma) = \frac{\sum_{j=1}^n \overline{f_j(\varsigma)} g_j(z,\varsigma)}{\|f(\varsigma)\|^2}$$

With this notation in place, we can state the following result.

Proposition 4.4. *For any integer $N \le n$ there are three differential forms K_N, H_N, and L_N of type $(n, n-1), (n-1, n-2), (n-1, n-1)$ in the variable ς, respectively, such that for any $z \in U$ and any $\theta \in \mathcal{D}(U)$, $\theta \equiv 1$ in a neighborhood of V, one has*

$$h(z) = \sum_{|k| \le N-n} < \bar\partial \frac{1}{f^{*(k+\underline{1})}}(\varsigma),\, \theta(\varsigma)\Theta(z,\varsigma) > f(z)^{*k}$$

$$+ \int_{\partial U} (\Gamma_h(z,\varsigma) K_N(z,\varsigma) + H_N(z,\varsigma) \wedge \bar\partial T(z,\varsigma) + L_N(z,\varsigma) \wedge T(z,\varsigma))$$

$$(4.14)$$

The differential forms are given by

$$K_N = \frac{1}{(2\pi i)^n} \sum_{\alpha_0+\alpha_1=n-1} \binom{N}{\alpha_1} \left(\frac{\sum_{j=1}^n \overline{f_j(\varsigma)} f_j(z)}{\|f(\varsigma)\|^2} \right)^{N-\alpha_1} \frac{S \wedge \bar\partial S^{\alpha_0} \wedge \bar\partial A^{\alpha_1}}{\|z-\varsigma\|^{2(\alpha_0+1)}}$$

$$H_N = \frac{1}{(2\pi i)^n} \sum_{\alpha_0+\alpha_1=n-2} \binom{N}{\alpha_1} \left(\frac{\sum_{j=1}^n \overline{f_j} f_j(z)}{\|f\|^2} \right)^{N-\alpha_1} \frac{S \wedge \bar\partial S^{\alpha_0} \wedge \bar\partial A^{\alpha_1}}{\|z-\varsigma\|^{\alpha_0+1}}$$

$$L_N = \frac{1}{(2\pi i)^n} \binom{N}{n-1} \left(\frac{\sum_{j=1}^n \overline{f_j} f_j(z)}{\|f\|^2} \right)^{N-n+1} \bar\partial A^{n-1}.$$

Proof. We use Proposition 2.8 and its proof, with the two weights introduced in the preceding proof of Proposition 4.3. Because of $q^{(x)}$, there is no boundary term in the representation formula of Proposition 2.8. Let us study the integral of the first half of the kernel $\tilde{P}_h$, i.e.,

$$\frac{1}{(2\pi i)^n} \int_U \Gamma_h(z,\varsigma) \sum_{\alpha_0+\alpha_1=n} \frac{\Gamma_0^{(\alpha_0)} \Gamma_1^{(\alpha_1)}}{\alpha_0! \alpha_1!} (\bar\partial Q^{(x)})^{\alpha_0} \wedge \bar\partial Q_\mu^{\alpha_1} ,$$

where

$$\frac{\Gamma_0^{(\alpha_0)}}{\alpha_0!} = \binom{n+1}{\alpha_0} \chi(\varsigma)^{n+1-\alpha_0}$$

$$\frac{\Gamma_1^{(\alpha_1)}}{\alpha_1!}(z,\varsigma;\mu) = \binom{N}{\alpha_1} \left[1 - |f(\varsigma)|^{*\mu 2} + |f(\varsigma)|^{*\mu 2} \frac{\sum_{j=1}^n \overline{f_j(\varsigma)} f_j(z)}{\|f(\varsigma)\|^2} \right]^{N-\alpha_1}$$

These terms have already been studied in the proof of Proposition 4.3, and, when we let first $\mu = 0$ and then $\chi \to \chi_U$, we get

$$\sum_{|k| \leq N-n} < \bar{\partial} \frac{1}{f^{*(k+\underline{1})}}, \Gamma_h \theta \bigwedge_{j=1}^{n} g_j > f(z)^{*k}$$

$$+ \frac{1}{(2\pi i)^n} \int_{\partial U} \Gamma_h \sum_{\alpha_0 + \alpha_1 = n-1} \binom{N}{\alpha_1} \left(\frac{\sum_{j=1}^{n} \overline{f_j} f_j(z)}{\|f\|^2} \right)^{N - \alpha_1} \frac{S \wedge \bar{\partial} S^{\alpha_0} \wedge \bar{\partial} A^{\alpha_1}}{\|z - \zeta\|^{2(\alpha_0 + 1)}}$$

$$= \sum_{|k| \leq N-n} < \bar{\partial} \frac{1}{f^{*(k+\underline{1})}}, \Gamma_h \theta \bigwedge_{j=1}^{n} g_j > f(z)^{*k} + \int_{\partial U} \Gamma_h(z, \zeta) K_N(z, \zeta)$$

$$(4.15)$$

Let us consider now the integral corresponding to the second half of $\tilde{P}_h$. It can be written as

$$\frac{1}{(2\pi i)^n} \int_U \sum_{\alpha_0 + \alpha_1 = n-1} \frac{\Gamma_0^{(\alpha_0)} \Gamma_1^{(\alpha_1)}}{\alpha_0! \alpha_1!} (\bar{\partial} Q^{(\chi)})^{\alpha_0} \wedge \bar{\partial} Q_\mu^{\alpha_1} \wedge \bar{\partial} T , \qquad (4.16)$$

where T is the differential form associated to τ. In this sum we have two parts, one corresponds to $\alpha_0 = 0$, and the other to the remaining terms. We study first the latter. The behaviour is exactly the same as that of the second integral in (4.5). One obtains, after replacing $\alpha_0 - 1$ by α_0 as in the formula (2.15), the expression

$$\frac{1}{(2\pi i)^n} \int_{\partial U} \sum_{\alpha_0 + \alpha_1 = n-2} \binom{N}{\alpha_1} \left(\frac{\sum_{j=1}^{n} \overline{f_j} f_j(z)}{\|f\|^2} \right)^{N - \alpha_1} \frac{S \wedge \bar{\partial} S^{\alpha_0} \wedge \bar{\partial} A^{\alpha_1} \wedge \bar{\partial} T}{\|z - \zeta\|^{2(\alpha_0 + 1)}}$$

$$= \int_{\partial U} H_N(z, \zeta) \wedge \bar{\partial} T . \qquad (4.17)$$

Finally, we are left with the term of (4.16) with $\alpha_0 = 0$, $\alpha_1 = n - 1$,

$$\frac{1}{(2\pi i)^n} \binom{N}{n-1} \int_U \chi^{n+1} \left[1 - |f|^{*\mu 2} + |f|^{*\mu 2} \frac{\sum_{j=1}^{n} \overline{f_j} f_j(z)}{\|f\|^2} \right]^{N-n+1} \bar{\partial} Q_\mu^{n-1} \wedge \bar{\partial} T .$$

This integral can be transformed using Stokes's theorem into

$$\frac{-1}{(2\pi i)^n} \binom{N}{n-1} \int_U \chi^{n+1} \bar{\partial} \left[1 - |f|^{*\mu 2} + |f|^{*\mu 2} \frac{\sum_{j=1}^{n} \overline{f_j} f_j(z)}{\|f\|^2} \right]^{N-n+1} \wedge \bar{\partial} Q_\mu^{n-1} \wedge T$$

$$+ \frac{-(n+1)}{(2\pi i)^n} \binom{N}{n-1} \int_U \left[1 - |f|^{*\mu 2} + |f|^{*\mu 2} \frac{\sum_{j=1}^{n} \overline{f_j} f_j(z)}{\|f\|^2} \right]^{N-n+1} \chi^n \bar{\partial} \chi \wedge \bar{\partial} Q_\mu^{n-1} \wedge T .$$

$$(4.18)$$

The second integral in the right-hand side of (4.18), can be studied as that in (4.17). Since there are no points of V near ∂U, we can let $\mu = 0$ and, after that, let $\chi \to \chi_U$. Lemma 2.5 is applicable and shows this integral converges to

$$\frac{1}{(2\pi i)^n} \binom{N}{n-1} \int_{\partial U} \left[\frac{\sum_{j=1}^n \overline{f_j} f_j(z)}{\|f\|^2} \right]^{N-n+1} \bar{\partial} A^{n-1} \wedge T$$

$$= \int_{\partial U} L_N(z, \zeta) \wedge T. \tag{4.19}$$

We now have to compute the contribution of the first integral in the right-hand side of (4.18). Let us rewrite

$$\Phi_\mu(z, \zeta) := 1 - |f|^{*\mu 2} + |f|^{*\mu 2} \frac{\sum_{j=1}^n \overline{f_j} f_j(z)}{\|f\|^2} = 1 - |f|^{*\mu 2} + \sum_{j=1}^n \eta_{j,\mu} f_j(z)$$

$$\eta_{j,\mu} := |f|^{*\mu 2} \frac{\overline{f_j}}{\|f\|^2} .$$

Then,

$$-\bar{\partial} \Phi_\mu \wedge \bar{\partial} Q_\mu^{n-1} = (-1)^{\frac{(n-2)(n-1)}{2}} (n-1)! \left(\bigwedge_{j=1}^n \bar{\partial} \eta_{j,\mu} \right) \wedge \left(\sum_{j=1}^n (-1)^j (f_j(z) - f_j(\zeta)) g_{[j]} \right)$$

$$= (-1)^{\frac{(n-2)(n-1)}{2}} n! \mu \frac{|f|^{*n\mu 2}}{\|f\|^{2n}} \overline{\partial f} \wedge \left(\sum_{j=1}^n (-1)^j (f_j(z) - f_j(\zeta)) g_{[j]} \right). \tag{4.20}$$

To conclude the proof, we take the analytic continuation of (4.20) to $\mu = 0$, and, in the same way we got to (4.15), we deduce from Proposition 3.16 that the contribution of (4.20) is

$$\sum_{|k| \le N-n} (-1)^{n-1} < \bar{\partial} \frac{1}{f^{*(k+\underline{1})}}, \theta \left(\sum_{j=1}^n (-1)^j (f_j(z) - f_j(\zeta)) g_{[j]} \right) \wedge T > f(z)^{*k}. \tag{4.21}$$

Adding (4.15) to (4.21), one gets the statement of the proposition. $\qquad\square$

The representation formulas of the last two propositions, with convenient choices of weights, will provide explicit division formulas in certain algebras of entire functions. In fact, they will allow us the extend local division formulas, i.e., near V, to global ones. Let us point out that the Cauchy-Weil formula itself can be used to solve division problems.

When $f_1, \ldots, f_n$ define a discrete variety in U, the Cauchy-Weil formula provides a division formula with a remainder term inside any Weil polyhedron Δ_ϵ of the form $\{|f_1| \leq \epsilon_1, \ldots, |f_n| \leq \epsilon_n\}$. For any $h \in H(\Delta_\epsilon) \cap C(\bar{\Delta}_\epsilon)$, we have that for $z \in \Delta_\epsilon$

$$h(z) = \frac{1}{(2\pi i)^n} \int_{\Gamma_f(\epsilon)} \frac{h(\zeta) g(z, \zeta)}{(f_1(\zeta) - f_1(z)) \cdots (f_n(\zeta) - f_n(z))} \qquad (4.22)$$

Formula (4.22) can be rewritten as

$$h(z) = \sum_{k=1}^{n} \sum_{\substack{S \subset \{1, \ldots, n\} \\ \#S = k}} F_S(z) A_S(z) + \frac{1}{(2\pi i)^n} \int_{\Gamma_f(\epsilon)} \frac{h(\zeta) g(z, \zeta)}{f_1(\zeta) \cdots f_n(\zeta)} \qquad (4.23)$$

where we have denoted, for any subset S of $\{1, \ldots, n\}$,

$$F_S(z) = \prod_{j \in S} f_j(z)$$

$$A_S(z) = \frac{1}{(2\pi i)^n} \int_{\Gamma_f(\epsilon)} \frac{h(\zeta) \prod_{j \notin S} (f_j(\zeta) - f_j(z))}{\prod_{j=1}^{n} f_j(\zeta)(f_j(\zeta) - f_j(z))} g(z, \zeta).$$

This expression is just obtained by developping

$$\prod_{j=1}^{n} (f_j(\zeta))$$

as

$$\prod_{j=1}^{n} (f_j(z) + (f_j(\zeta) - f_j(z)))$$

(for more detailed computations, see [4, §1, 194-195].) The remainder term in formula (4.23) can be expressed as

$$< \bar{\partial} \frac{1}{f}(\zeta), \theta(\zeta) h(\zeta) g(z, \zeta) >,$$

where θ denotes any test function with support in Δ, identically equal to one near the set of common zeros of $f_1, \ldots, f_n$. Since the residue current is annihilated by the ideal generated by $f_1, \ldots, f_n$, the remainder term in the division formula (4.23) is zero when h is in the ideal $I_{loc}(f_1, \ldots, f_n)$.

Remark 4.5. With the methods just developed, it is easy to give another proof of the Local Duality Theorem (Theorem 3.21.). This follows from the fact that the Cauchy-Weil formula furnishes an explicit division formula with remainder even in the case when the number of functions p is smaller than n. Assume, for example, that $f_1, \ldots, f_p$ define a system in normal position in a ball U about the origin, $f_j(0) = 0$, and let $f_{p+1}, \ldots, f_n$ be linear functions such that $f_1, \ldots, f_n$ define the origin as an isolated common zero. In a Weil polyhedron Δ_ϵ for the whole system, we have written above an explicit division formula for any h belonging to the local ideal $I_{loc}(f_1, \ldots, f_n)$. In particular, one can use this formula for any $h \in I_{loc}(f_1, \ldots, f_p)$. What we have to verify is that formula (4.23) can be rearranged so that it gives a division formula involving only f_j for $1 \le j \le p$. One uses a method of Berenstein and Taylor (reference supra.) We start by rewriting the Cauchy-Weil formula:

$$h(z) = \frac{1}{(2\pi i)^n} \int_{\Gamma_f(\epsilon)} \left(\frac{1}{\prod_{1 \le j \le n} (f_j(\zeta) - f_j(z))} - \frac{1}{\prod_{1 \le j \le n} f_j(\zeta)} \right) h(\zeta) g(z, \zeta). \quad (4.24)$$

This follows from the fact that the subtrahend is zero, since h is in the ideal. In order to simplify a bit the writing of the formulas we let

$$H(z, \zeta) := \frac{1}{\prod_{1 \le j \le n} (f_j(\zeta) - f_j(z))} - \frac{1}{\prod_{1 \le j \le n} f_j(\zeta)}$$

$R(\zeta) := \prod_{p+1 \le j \le n} f_j(\zeta)$, and $\varphi_j(z, \zeta) := f_j(\zeta) - f_j(z)$. By analogy with the notation used in the proof of (4.23), if $\mathcal{T} \subset \{1, \ldots, n\}$ we denote $\varphi_{\mathcal{T}} := \prod_{j \in \mathcal{T}} \varphi_j$. We can now rewrite H using the identity

$$H(z, \zeta) \prod_{j=1}^{n} f_j(\zeta) \varphi_j(z, \zeta) = R(\zeta) \prod_{j=1}^{p} (\varphi_j(z, \zeta) + f_j(z)) - \prod_{j=1}^{n} \varphi_j(z, \zeta). \quad (4.25)$$

We now develop the right-hand side of (4.25) as

$$R(\zeta) \left(\sum_{j=1}^{p} \sum_{\substack{S \subseteq \{1, \ldots, p\} \\ \#S = j}} F_S(z) \varphi_S'(z, \zeta) \right) - \varphi_{\{1, \ldots, p\}}(z, \zeta)(\varphi_{\{p+1, \ldots, n\}}(z, \zeta) - R(\zeta)),$$

$$(4.26)$$

with $S' = \{1, \ldots, p\} \setminus S$. We replace H in (4.24) by the expression obtained from (4.25) and (4.26). We obtain

$$h(z) = \frac{1}{(2\pi i)^n} \sum_{j=1}^{p} \sum_{\substack{S \subseteq \{1, \ldots, p\} \\ \#S = j}} \left(\int_{\Gamma_f(\epsilon)} \frac{h(\zeta) R(\zeta) \varphi_{S'}(z, \zeta) g(z, \zeta)}{\prod_{1 \le j \le n} f_j(\zeta) \varphi_j(z, \zeta)} \right) F_S(z)$$

$$+ \frac{1}{(2\pi i)^n} \int_{\Gamma_f(\epsilon)} \frac{h(\zeta) \varphi_{\{1, \ldots, p\}}(z, \zeta)(R(\zeta) - \varphi_{\{p+1, \ldots, n\}}(z, \zeta)) g(z, \zeta)}{\prod_{j=1}^{n} f_j(\zeta) \varphi_j(z, \zeta)},$$

$$(4.27)$$

We claim that the second term of (4.27) vanishes. By linearity, it is enough to show this is true when $h = \tilde{h} f_1$ in $H(U)$. In this case, the factor f_1 disappears from the denominator, so that the integrand becomes an $(n,0)$ holomorphic form in a neighborhood of the set $\{|f_1| \leq \epsilon_1, |f_j| = \epsilon_j, 2 \leq j \leq n\}$, and an application of Stokes's theorem shows that the second term of (4.27) is indeed zero. (We have already used an argument like this one in the proof of Proposition 3.12.)

Remark 4.6. As a corollary of (4.23) we can give a different proof of the Briançon-Skoda theorem [11]. This proof is based on the fact that we can test the membership in an ideal using a duality method. The argument can be adapted, as we have mentioned already in Chapter 3, §5, to a purely algebraic context, which is not the case for the original proof of Briançon and Skoda or in the proof we gave earlier. Since we intend to use this idea repeatedly in the rest of the book, we sketch this proof of the theorem of Briançon-Skoda in the discrete, complete intersection case.

Let us show that if $f_1, \ldots, f_n$ are holomorphic functions in U defining a zero dimensional variety V, and if $h \in H(U)$ satisfies in any compact subset K of U an inequality of the form

$$|h(\zeta)| \leq C_K \|f(\zeta)\| \quad \forall \zeta \in K, \tag{4.28}$$

then $h^n \in I_{loc}(f_1, \ldots, f_n)$. We can show that for any $\phi \in \mathcal{D}(U)$, which is holomorphic near V,

$$< \bar{\partial}\frac{1}{f}, h^n \phi d\zeta >= 0$$

This can be seen as follows.

$$< \bar{\partial}\frac{1}{f}, h^n \phi d\zeta >= \lim_{k \to \infty} \frac{1}{(2\pi i)^n} \int_{\Gamma_f(\epsilon(k))} \frac{h^n(\zeta)\phi(\zeta)}{f_1(\zeta) \cdots f_n(\zeta)} d\zeta,$$

where $\epsilon(k)$ tends to 0 over a sequence of non-critical values for $(|f_1|^2, \ldots, |f_n|^2)$, in such a way that every component $\epsilon_j(k) \sim \frac{1}{k}$. Furthermore, from the Coarea formula we can assume that

$$\lim_{k \to \infty} \int_{\Gamma_f(\epsilon(k))} |d\zeta_1 \wedge \ldots \wedge d\zeta_n| = 0.$$

Therefore, as $\epsilon_1(k) \cdots \epsilon_n(k) \sim k^{-n}$, it follows from (4.28) that

$$\max_{\zeta \in \Gamma_f(\epsilon(k))} |h(\zeta)^n| \leq Ck^{-n}.$$

Thus,

$$
\left| \int_{\Gamma_f(\epsilon(k))} \frac{h^n \phi}{f_1 \cdots f_n} \right| \leq \frac{\|\phi\|_\infty}{\epsilon_1(k) \cdots \epsilon_n(k)} \Big(\max_{\Gamma_f(\epsilon(k))} |h^n| \Big) \int_{\Gamma_f(\epsilon(k))} |d\zeta_1 \wedge \ldots \wedge d\zeta_n|
$$

$$
\leq C' \int_{\Gamma_f(\epsilon(k))} |d\zeta_1 \wedge \ldots \wedge d\zeta_n| \longrightarrow 0.
$$

So we have that $< \bar{\partial}\frac{1}{f}, h^n \phi d\zeta > = 0$ for any $\phi \in \mathcal{D}(U)$, holomorphic near V. From (4.23) in a Weil polyhedra surrounding the set V, we conclude that $h^n \in I_{loc}(f_1, \ldots, f_n)$.

§2 The Grothendieck residue in the discrete case

As we have seen in the last section, the Grothendieck residue in the discrete case is closely related to the Cauchy-Weil formula; we shall develop here some of its important properties. The first important one concerns the order of the residue current in the discrete case, the second one is the law of transformation which will enable us to use the residue as a computational tool in Chapter 5.

Let $f_1, \ldots, f_n$ be holomorphic functions in a neighborhood U of the origin in $\mathbb{C}^n$ such that $\{z \in U : f_j(z) = 0 \ \forall j\} = \{0\}$. Let D be a ball $B(0, \rho) \subset\subset U$. Consider the integral

$$
N(f) := (-1)^{\frac{n(n-1)}{2}} \frac{(n-1)!}{(2\pi i)^n} \int_{\partial D} \frac{\sum_{k=1}^n (-1)^{k-1} \bar{f}_k \overline{df_{[k]}} \wedge df}{\|f\|^{2n}} . \tag{4.29}
$$

We have seen in Corollary 2.11 that $N(f)$ is an integer and that, if $\varphi_1, \ldots, \varphi_n$ are holomorphic functions in U such that $\|f - \varphi\|(\zeta) < \|f\|(\zeta)$ for every $\zeta \in \partial D$, then $N(f) = N(\varphi)$. In particular, this is true when $\varphi = (f_1 - \epsilon_1, \ldots, f_n - \epsilon_n)$, with $\|\epsilon\| < \min_{\partial D} \|f\|$. If, additionally, ϵ is not a critical value of the map f, then $N(f - \epsilon)$ represents the number of distinct common zeroes of $f_j - \epsilon_j$ in $\bar{D}$, all of them being simple zeroes. Recall that $N(f)$ is known as the multiplicity of f at the origin. One can show (see [7, p. 616]) that the multiplicity equals the topological degree of the map f at the origin, it can also be interpreted as the dimension of the quotient vector space

$$
\frac{{}_n\mathcal{O}_0}{\mathcal{I}} \qquad \text{where} \quad \mathcal{I} := (f_1, \cdots, f_n) \, {}_n\mathcal{O}_0 .
$$

The multiplicity provides a bound for the exponent of the local Nullstellensatz at the origin, as shown by the following proposition.

Proposition 4.6. *Let $f_1, \ldots, f_n$ be holomorphic functions defining the origin as an isolated zero, with multiplicity ν. Let h be an holomorphic function that $h(0) = 0$. Then*

$$h^\nu \in (f_1, \cdots, f_n)_n \mathcal{O}_0.$$

Proof. Let $\mathcal{G}$ be the graph of f in $\mathbb{C}^{2n}$. For some balls $D' = B(0, r')$ and $D'' = B(0, r'')$ in $\mathbb{C}^n$, we have

$$\mathcal{G} \cap (D' \times D'') = \{(\zeta, w) \in D' \times D'' : w_1 = f_1(\zeta), \ldots, w_n = f_n(\zeta)\}$$

and the projection on the second factor π is such that its restriction to $\mathcal{G}$ is a branched covering of D'' with μ sheets [7, p. 667]. For $w \in D''$, let $f^{-1}(w) = \{\zeta^{(1)}(w), \ldots, \zeta^{(\nu)}(w)\}$, the ordering of these elements is irrelevant here. Define the function

$$H(t, w) := \prod_{j=1}^{\nu} (t - h(\zeta^{(j)}(w))).$$

Corollary 2.10 shows that the Newton sums of the $h(\zeta^{(j)}(w))$ are holomorphic funtions of w, thus, their symmetric functions, and, hence, the function $(t, w) \mapsto H(t, w)$ is holomorphic. Therefore, the function $\phi(\zeta, w) := H(h(\zeta), w)$ is holomorphic in $D' \times D''$. Note that $\phi(\zeta, 0) = h(\zeta)^\nu$. The function ϕ is zero on $\mathcal{G}$ and, since $\mathcal{G}$ is clearly a submanifold of $\Delta' \times \Delta''$, there exist Hefer functions $\alpha_1, \ldots, \alpha_n$ in $\Delta' \times \Delta''$ such that

$$\phi(\zeta, w) = \sum_{j=1}^{n} \alpha_j(\zeta, w)(f_j(\zeta) - w_j). \tag{4.30}$$

(See proof of Theorem 3.21.) Consider (4.30) when $w = 0$, one obtains

$$h(\zeta)^\nu = \sum_{j=1}^{n} \alpha_j(\zeta, 0) f_j(\zeta).$$

This concludes the proof of Proposition 4.6.   □

We can illustrate this proposition by proving a variant of it in an algebraic setting.

Proposition 4.7. *Let $P_1, \ldots, P_m$ be polynomials in $\mathbb{C}[X_1, \ldots, X_n]$, $m \leq n$, which define a germ of a complete intersection variety V at the origin, and let $\mathcal{I}$ be the ideal they generate in $_n\mathcal{O}_0$. For any function h, holomorphic in a neighborhood of the origin, and vanishing on V, one has*

$$h^k \in \mathcal{I} \quad \text{for} \quad k = \deg(P_1) \cdots \deg(P_m).$$

Proof. The proof is similar to the proof of Proposition 4.6. After a rotation of coordinates and restriction to a neighborhood of the origin, we can assume that $\{P_1 = \ldots = P_m = \zeta_{m+1} = \ldots = \zeta_n = 0\} = \{0\}$. Then, we consider the application $\tilde{P}$ from $\mathbb{C}^n$ to $\mathbb{C}^n$ defined by

$$\tilde{P}(\zeta) = (\zeta_{m+1}, \ldots, \zeta_n, P_1(\zeta), \ldots, P_m(\zeta)),$$

and we denote, as before, $\mathcal{G}$ the graph of P in $\mathbb{C}^{n+m}$. Let π be the projection from $\mathcal{G}$ onto the last n coordinates, so that $\pi(\zeta, P(\zeta)) = \tilde{P}(\zeta)$. As 0 is an isolated point in $\pi^{-1}(0)$, there are balls $D' \subset \mathbb{C}^n$, $D'' \subset \mathbb{C}^m$ such that, the restriction $\pi : \mathcal{G} \cap D' \times D'' \to D''$ is a branched covering with ν sheets. Here ν is the multiplicity of the origin as a zero of the polynomial map $\tilde{P}$, and it is bounded by $\deg(P_1)\cdots\deg(P_m)$, as a consequence of the inequality (2.27), which followed Bezout's Theorem 2.22. As before, for $\zeta'' \in D''$, we introduce $\pi^{-1}(\zeta'')$, which is a finite set of cardinal ν, $\{\zeta^{(1)}, \ldots, \zeta^{(\nu)}\}$. Let $\widetilde{\zeta^{(j)}}$, $j = 1, \ldots, \nu$, be the projections of the ν points of $\pi^{-1}(\zeta'')$ on the space of their n first coordinates. Then, as in the preceding proof, one can define a holomorphic function H on $\mathbb{C} \times D''$ by

$$H(t, \zeta'') = \prod_{j=1}^{\nu} (t - \widetilde{\zeta^{(j)}}).$$

We now introduce the following element in $_{n+m}\mathcal{O}_0$,

$$(\zeta, w) \mapsto \phi(\zeta, w) = H(h(\zeta), \zeta_{m+1}, \ldots, \zeta_n, w_1, \ldots, w_m).$$

One can see at once that ϕ vanishes on the germ of the set $\mathcal{G}$. We conclude the argument as in the proof of the last proposition. □

Let us point out here a consequence of Proposition 4.7 that we will develop further in Chapter 5. Let $P_1, \ldots, P_m$ be polynomials in $\mathbb{C}[X_1, \ldots, X_n]$ such that the corresponding homogeneous polynomials ${}^hP_1, \ldots, {}^hP_n$ define locally a complete intersection in $\mathbb{C}^{n+1}$. Recall that the homogeneous polynomial ${}^hP$ in $\mathbb{C}[X_0, \ldots, X_n]$ corresponding to a polynomial $P \in \mathbb{C}[X_1, \ldots, X_n]$ is given by

$${}^hP(X_0, \ldots, X_n) = X_0^{\deg(P)} P(X_1/X_0, \ldots X_n/X_0) \quad (X_0 \neq 0).$$

Let $Q \in \mathbb{C}[X_1, \ldots, X_n]$ vanish at all the common zeroes of $P_1, \ldots, P_m$, then, the homogeneous polynomial $X_0\,{}^hQ$, vanishes on the common zeroes of ${}^hP_1, \ldots, {}^hP_n$ in $\mathbb{C}^{n+1}$, thus, we can apply Proposition 4.7 to obtain

$$(X_0\,{}^hQ)^{\deg(P_1)\cdots\deg(P_m)} \in I({}^hP_1, \ldots, {}^hP_m).$$

Hence, there are homogeneous polynomials $\tilde{A}_1, \ldots, \tilde{A}_m$ such that

$$(X_0{}^h Q)^{\deg(P_1)\cdots\deg(P_m)} = \tilde{A}_1{}^h P_1 + \ldots + \tilde{A}_m{}^h P_m$$
$$\text{and} \quad \deg(\tilde{A}_j{}^h P_j) = \deg(P_1)\cdots\deg(P_m)(\deg({}^h Q)+1). \qquad (4.31)$$

If we let $X_0 = 1$ in this identity, we obtain polynomials A_j of n variables, such that

$$Q^{\deg(P_1)\cdots\deg(P_m)} = A_1 P_1 + \ldots + A_n P_m ,$$
$$\deg(A_j P_j) \le (1+\deg(Q))\deg(P_1)\cdots\deg(P_m). \qquad (4.32)$$

This is a version of the algebraic Nullstellensatz that we will generalize in the next chapter. In fact, under our hypotheses, the exponent ν we gave is far away from being optimal, as we will see later. The result remains essentially true if we only assume that $P_1, \ldots, P_m$ define a complete intersection in $\mathbb{C}^n$ and, in that case, the exponent ν equal to the product of the degrees, cannot in general be sharpened. For more details see [10],[14], and also Chapter 5 below.

The multiplicity governs the order of the residue current in the discrete case.

Proposition 4.8. *Let $f_1, \ldots, f_n$ be functions holomorphic in some neighborhood U of $0 \in \mathbb{C}^n$, such that the origin is the only zero of the map f in U. Let ν denote the multiplicity of f at the origin, for $k \in \mathbb{N}^n$, $|k| < \nu$, there are complex constants c_k, so that, for any $\phi \in \mathcal{D}(U)$,*

$$< \bar{\partial}\frac{1}{f},\, \phi d\zeta >= \sum_{|k|<\nu} c_k \frac{\partial^{|k|}}{\partial\zeta^k}\phi(0). \qquad (4.33)$$

Proof. Proposition 3.27 shows that the residue current can be represented by a distribution with support at the origin, whence a finite linear combination of derivatives of the Dirac mass δ at 0.

Let us show now that, as we have seen earlier in the case of one variable, the only derivatives that appear are holomorphic derivatives, i.e., only derivatives with respect to the ζ_j. This is equivalent to the fact that the residue current $\bar{\partial}(1/f)$ is annihilated by the functions $\bar{\zeta}_j$. Recall that the action of the residue current on a test form ϕ can be obtained through our method of analytic continuation as the value at $\mu = 0$ of the map

$$\mu \longrightarrow \frac{1}{(2\pi i)^n}\mu^n \int \frac{|f|^{*\mu 2}}{f^{*\underline{1}}}\frac{\overline{\partial f}}{\overline{f^{*\underline{1}}}} \wedge \phi \qquad (4.34)$$

Of course, we are only interested in the case where, let us say, $\phi = \bar{\zeta}_1\psi$. As in the proof of Theorem 3.25, introduce a resolution of singularities, $(\mathcal{X}, \pi)$, for the hypersurface $\{f_1 \cdots f_n \zeta_1 \cdots \zeta_n = 0\}$. Let us consider a chart where, in a system of coordinates w centered at the origin, all the functions $\pi^*(f_j)$ and $\pi^*(\zeta_j)$, are (up to invertible holomorphic functions) monomials in w,

$$\pi^*(f_j) = u_j w^{\alpha_j}, \quad \pi^*(\zeta_j) = v_j w^{\beta_j} .$$

So that (4.34) becomes a linear combination of expressions of the type

$$\mu^n \int \pi^*(\bar\zeta_1)\pi^*\left(\frac{|f|^{*\mu 2}}{f^{*\underline{1}}}\right)\pi^*\left(\frac{\overline{\partial f}}{\overline{f^{*\underline{1}}}}\right) \wedge \Theta(w) \tag{4.35}$$

Θ is test form in the w variables. In order to study such an expression, we introduce the toroidal manifold $(\tilde{\mathcal{X}}, \tilde\pi)$ corresponding to the Newton polyedron $\Gamma^+(\alpha_1, \ldots, \alpha_n)$ and we transform (4.35) into a linear combination of expressions of the form

$$\mu^n \int \tilde\pi^*\pi^*(\bar\zeta_1)\tilde\pi^*\pi^*\left(\frac{|f|^{*\mu 2}}{f^{*\underline{1}}}\right)\tilde\pi^*\pi^*\left(\frac{\overline{\partial f}}{\overline{f^{*\underline{1}}}}\right) \wedge \theta(t) \tag{4.36}$$

where θ is a test form in the variables t corresponding to some local chart on $\tilde{\mathcal{X}}$. We will call m the distinguished monomial corresponding to this chart. Recall from the basic properties of toroidal desingularizations that one of the f_j, let say f_1, is such that $\tilde\pi^*\pi^*(f_1)(t) = y_1(t)m(t)$ and that m divides all the functions $\tilde\pi^*\pi^*(f_j)$, $j = 1, \ldots, n$. We now remark that one can write

$$\tilde\pi^*\pi^*\left(\frac{\overline{\partial f}}{\overline{f^{*\underline{1}}}}\right) = \left(\frac{\overline{\partial m}}{\overline{m}} + \frac{\overline{\partial y_1}}{\overline{y_1}}\right) \wedge S(t), \tag{4.37}$$

where S is a linear combination of differential forms of the type

$$\frac{s(t)}{\prod_{j \in J} \bar t_j},$$

with s a smooth form and J some subset of $\{1, \ldots, n\}$ of cardinal strictly smaller than n.

We claim that any t_j appearing in the distinguished monomial m, also appears in $\tilde\pi^*(w^{\beta_1})$. If we knew this, all the antiholomorphic singularities that arise from $\overline{\partial m}/\overline{m}$, which are all of the first order, would be compensated by the corresponding t_j appearing in $\tilde\pi^*(w^{\beta_1})$, hence, one would have at most $n - 1$ coordinates $\bar t_j$ in the denominator, and the proof of Proposition 3.6 shows that the value at $\mu = 0$ of the analytic continuation of (4.36) is zero. To prove the claim, we go back to our original coordinates, use the local Nullstellensatz at the origin (Proposition 4.6), which guarantees that ζ_1^ν belongs to $I_{loc}(f_1, \ldots, f_n)$. Since, in our local chart, m divides all the $\tilde\pi^*\pi^*(f_j)$, it divides $(\tilde\pi^*(w^{\beta_1}))^\nu$. This fact clearly implies that every t_j dividing m divides $\tilde\pi^*(w^{\beta_1})$, as claimed.

We have just shown that the residue current $\bar{\partial}(1/f)$ is killed by any multiple of every $\bar{\zeta}_j$, this immediately implies that the distribution representing the current contains only holomorphic derivatives of δ. To conclude the proof, note that if $\mathcal{M} = I_{loc}(\zeta_1, \ldots, \zeta_n)$ denotes the maximal ideal of $_n\mathcal{O}_0$, then the local Nullstellensatz implies that $\mathcal{M}^\nu \subseteq \mathcal{I}$. The local duality theorem (Theorem 3.21) implies that the order of the derivatives is at most $\nu - 1$. $\qquad\square$

From the computational point of view, one of the most important properties of the Grothendieck residue in the discrete case is the transformation law. It is in fact, quite often, the only tool we have at our disposal to compute the action of the residue current.

Theorem 4.9. *Let $f_1, \ldots, f_n, \varphi_1, \ldots, \varphi_n$ be $2n$ holomorphic functions in a neighborhood U of 0 in $\mathbb{C}^n$, such that 0 is the only zero in U of both the maps f and φ. Suppose that for $j = 1, \ldots, n$ we have*

$$\varphi_j = \sum_{k=1}^n a_{j,k}(\zeta) f_k(\zeta) \tag{4.38}$$

for some functions $a_{j,k}$ holomorphic in U. Then, if Δ denotes the determinant of the matrix $[a_{j,k}]_{1 \le j, k \le n}$, we have, for any ϕ in $\mathcal{D}(U)$,

$$< \bar{\partial}\frac{1}{f}, \phi d\zeta > = < \bar{\partial}\frac{1}{\varphi}, \Delta\phi d\zeta > . \tag{4.39}$$

Proof. Our proof is inspired in that suggested in ([7], p. 657-659). We divide it here in four steps. We will denote as J_f and J_φ the Jacobians of the maps f and φ.

Step 1. It is clear from (4.33) that if two test forms, $\phi_1 d\zeta$ and $\phi_2 d\zeta$ are such that ϕ_1 and ϕ_2 have Taylor polynomials which coincide up to the order equal to the maximum of the mutiplicities of f and φ at the origin, then the residue currents associated to f and φ coincide on both test forms. Therefore, we can replace an arbitrary test form $\phi d\zeta$ by $P\chi d\zeta$, P a polynomial and χ a test function that is identically one near the origin. Now, as we have seen in the the proof of Proposition 4.8, the functions $\bar{\zeta}_1, \ldots, \bar{\zeta}_n$ annihilate the currents $\bar{\partial}(1/f)$ and $\bar{\partial}(1/\varphi)$. Thus, we can assume that P is a holomorphic polynomial. Hence, it is sufficient to prove (4.39) holds for test functions $\phi d\zeta$, with the coefficient ϕ holomorphic near 0. One can shrink U, if necessary, so that one can assume that ϕ is holomorphic in U.

Step 2. We prove the result when $\Delta(0) \ne 0$ and $J_f(0) \ne 0$. In that case, we have also $J_\varphi(0) \ne 0$. From Corollary 2.10 and formula (3.32) (applied when $\beta = \underline{0}$), it follows that

$$< \bar{\partial}\frac{1}{f}, \phi d\zeta > = \frac{\phi(0)}{J_f(0)}$$

Similarly,

$$< \bar{\partial}\frac{1}{\varphi}, \Delta\phi d\zeta >= \frac{\Delta(0)\phi(0)}{J_\varphi(0)} = \frac{\phi(0)}{J_f(0)}$$

which concludes the proof of formula (4.39) in that case.

Step 3. We now only assume $\Delta(0) \neq 0$. We use that, for almost all small $\epsilon = (\epsilon_1, \ldots, \epsilon_n)$, the functions $f^{(\epsilon)} := f_1 - \epsilon_1, \ldots, f_n - \epsilon_n$ define a manifold in U (i.e., all the point are simple.) This map will be called a good approximation of f. Of course, we may assume that Δ does not vanish on U. Then, from the invertibility of the matrix $[a_{j,k}(\zeta)]_{1 \leq j,k \leq n}$, we have

$$U \cap (f^{(\epsilon)})^{-1}(0) = U \cap (\varphi^{(\epsilon)})^{-1}(0),$$

where the perturbation $\varphi^{(\epsilon)}$ is defined by

$$\varphi_j^{(\epsilon)}(\zeta) = \sum_{k=1}^n a_{j,k}(\zeta) f_k^{(\epsilon)}(\zeta).$$

On the other hand, at any zero ζ_0 in U of $f^{(\epsilon)}$, that is at any common zero in U of $\varphi^{(\epsilon)}$ (these two maps have the same zeroes in U), we have

$$J_{\varphi^{(\epsilon)}}(\zeta_0) = \Delta(\zeta_0) J_{f^{(\epsilon)}}(\zeta_0),$$

so that $\varphi^{(\epsilon)}$ and $f^{(\epsilon)}$ have only simple zeroes in U. So, following the results obtained in the preceding steps, for small values of ϵ,

$$\sum_{\alpha \in (f^{(\epsilon)})^{-1}(0)} < \bar{\partial}\frac{1}{f^{(\epsilon)}}, \phi d\zeta >_\alpha = \sum_{\alpha \in (\varphi^{(\epsilon)})^{-1}(0)} < \bar{\partial}\frac{1}{\varphi^{(\epsilon)}}, \phi \Delta d\zeta >_\alpha \qquad (4.40)$$

where $<,>_\alpha$ denotes the action of the local residue at the point α. Note that (4.40) is valid because the zeroes of $f^{(\epsilon)}$ in U coincide exactly with the zeroes of $\varphi^{(\epsilon)}$ in U. If one applies formula (3.32) again, it is clear that if u is a n-valued holomorphic map defining a discrete variety in U (with no zeroes on the boundary) and if $u^{(\epsilon)}$ is a sequence of perturbations of u converging to u (when ϵ tends to 0) in the algebra $\mathcal{B}(U)$, then

$$< \bar{\partial}\frac{1}{u^{(\epsilon)}}, \phi d\zeta > \longrightarrow < \bar{\partial}\frac{1}{u}, \phi d\zeta >$$

when ϵ approaches 0. Formula (4.39) is shown to be correct in the present case, just take limits in (4.40) when ϵ tends to 0.

Step 4. We look now at the most general situation, that is when $\Delta(0) = 0$. We introduce first a continuous deformation of the matrix $\mathcal{A} = \mathcal{A}^{(0)} = [a_{j,k}]_{1 \leq j,k \leq n}$, namely, $(\mathcal{A}^{(\epsilon)})_\epsilon$, in such a way that for any $\epsilon \neq 0$, $\det(\mathcal{A}^{(\epsilon)}(0)) \neq 0$. We set

$$\varphi_j^{(\epsilon)}(\zeta) = \sum_{k=1}^n a_{j,k}^{(\epsilon)}(\zeta) f_k(\zeta).$$

If U' is a ball such that $\bar{U}' \subset U$, we can assume that for ϵ sufficiently small $(\varphi^{(\epsilon)})^{-1}(0) \cap \partial U' = \emptyset$. This implies, of course, that $(\varphi^{(\epsilon)})^{-1}(0) \cap U'$ is finite. Among the points of $(\varphi^{(\epsilon)})^{-1}(0)$ is the origin. From Step 3, we have

$$< \bar{\partial} \frac{1}{f}, \phi d\zeta >_0 = < \bar{\partial} \frac{1}{\varphi^{(\epsilon)}}, \phi \det(\mathcal{A}^{(\epsilon)}) d\zeta >_0 . \tag{4.41}$$

Let us now consider some point $\alpha \neq 0$, which is a zero of $\varphi^{(\epsilon)}$. In some neighborhood of α one can write, using Cramer's rule,

$$\det(\mathcal{A}^{(\epsilon)}) f_j(\zeta) = \sum_{k=1}^{n} A_{j,k} \varphi_k^{(\epsilon)}(\zeta).$$

Therefore, as the $f_1, \ldots, f_n$ do not vanish simultaneously at α, we can express $\det(\mathcal{A}^{(\epsilon)})$, in a neighborhood of α, as an element of the local ideal, at the point α, generated by $\varphi_1^{(\epsilon)}, \ldots, \varphi_n^{(\epsilon)}$. Using the local duality theorem,

$$< \bar{\partial} \frac{1}{\varphi^{(\epsilon)}}, \phi \det(\mathcal{A}^{(\epsilon)}) d\zeta >_\alpha = 0. \tag{4.42}$$

Adding (4.41) and (4.42), one has

$$< \bar{\partial} \frac{1}{f}, \phi d\zeta >_0 = \sum_{\alpha \in (\varphi^{(\epsilon)})^{-1}(0) \cap U'} < \bar{\partial} \frac{1}{\varphi^{(\epsilon)}}, \phi \det(\mathcal{A}^{(\epsilon)}) d\zeta >_\alpha . \tag{4.43}$$

The complete sum of residues on the right-hand side of (4.43), can be expressed using formula (3.32) as

$$\frac{(-1)^{\frac{(n-1)n}{2}} (n-1)!}{(2\pi i)^n} \int_{\partial U'} \Omega(\varphi^{(\epsilon)}, \zeta) \wedge \det(\mathcal{A}^{(\epsilon)}) \phi \, d\zeta$$

Letting $\epsilon \to 0$ we get the conclusion of the theorem. $\square$

Remark 4.10. What we have done above in Step 4 shows that the result is true in a global setting. Namely, if $f_1, \ldots, f_n$ and $\varphi_1, \ldots, \varphi_n$ define discrete varieties in some open set $U \subset \mathbb{C}^n$, and if there are holomorphic functions $a_{j,k}$ in U such that every $\varphi_j = \sum_{k=1}^{n} a_{j,k} f_k$ in U, then the transformation law (4.39) is valid for any $\phi \in \mathcal{D}(U)$.

§3 The Grothendieck residue in the algebraic case

As we have already mentioned in Remark 4.6, the Grothendieck residue, because of its dualizing properties, appears to play an important role in algebraic situations. The next proposition will show that most of the computations one can make, using the Grothendieck residue, keep track of the underlying algebraic structure. In this context, the Dolbeault point of view is more helpful than the currents point of view.

Let $\mathbb{F}$ be a number field and $P_1, \ldots, P_n$ be polynomials in $\mathbb{F}[X_1, \ldots, X_n]$ defining a discrete variety $V = \{P_1 = \ldots = P_n = 0\}$ in $\mathbb{C}^n$. Since V is a finite set and the residue current $\bar{\partial}(1/P)$ is supported by V, for any function ϕ, which is $\mathcal{C}^\infty$ in a neighborhood of V (but not of compact support), one can consider the sum of all the local residues and still denote it $< \bar{\partial}(1/P), \phi d\zeta >$. Note that one needs ϕ only to be differentiable of a finite order (namely, the maximum of the multiplicities at the points of V, i.e., at most the product of the degrees of the P_j.) We keep the above assumptions on the polynomials P_j throughout this section.

Proposition 4.11. *Let R be a rational function in $\mathbb{F}(X_1, \ldots, X_n)$, with no poles on V. Then*

$$< \bar{\partial}\frac{1}{P}, Rd\zeta >= \sum_{\alpha \in V} < \bar{\partial}\frac{1}{P}, Rd\zeta >_\alpha$$

is an element of F.

Proof. Let $R = G/H$, where G and H are coprime in $\mathbb{F}[X_1, \ldots, X_n]$, so that H does not vanish on V. Let $N = \deg(P_1) \cdots \deg(P_n)$. Recall that using elimination theory (see [16, vol. 2]) we can find n polynomials $Q_j \in \mathbb{F}[X_j]$ in the ideal I generated by $P_1, \ldots, P_n$ in $\mathbb{F}[X_1, \ldots, X_n]$, let us write

$$Q_j(X) = Q_j(X_j) = \sum_{k=1}^{n} a_{j,k} P_k(X), \quad \Delta = \det[a_{j,k}] \in \mathbb{F}[X_1, \ldots, X_n].$$

As a consequence of Dirichlet's pigeonhole principle, we can choose $\lambda_0, \ldots, \lambda_n$ in $\mathbb{Z}$ such that the polynomial

$$\tilde{H} = \lambda_0 H + \lambda_1 P_1^{n+1} + \ldots + \lambda_n P_n^{N+1}$$

does not vanish on $W = \{Q_1 = \ldots = Q_n = 0\}$ (see, for example, Lemma 5.2 in [5].) In fact, one can choose $|\lambda_k| \leq \#(W) \leq \deg(Q_1) \cdots \deg(Q_n)$. Proposition 4.8 and the choice of N ensures that

$$< \bar{\partial}\frac{1}{P}, Rd\zeta >=< \bar{\partial}\frac{1}{P}, \frac{G}{\tilde{H}}d\zeta > .$$

Using now the global version of the Transformation Law (Remark 4.10), we obtain

$$< \bar{\partial}\frac{1}{P}, \, Rd\zeta > = < \bar{\partial}\frac{1}{Q}, \, \Delta\frac{G}{\tilde{H}}d\zeta > . \tag{4.44}$$

Factorizing each Q_j into irreducible factors in $\mathbb{F}[X_j]$, we have

$$Q_j = Q_{j,1}^{n_{j,1}} \cdots Q_{j,s}^{n_{j,s}} ,$$

where $s = s(j)$. Choose $\epsilon \in (\mathbb{R}^+)^n$ such that the set $D_Q(\epsilon) = \{|Q_j(\zeta_j)| < \epsilon, \, j = 1, \ldots, n\}$ is a union of disjoint Weil polyhedra, and that the function $G/\tilde{H}$ is holomorphic in a neighborhood of $\bar{D}_Q(\epsilon)$. Thus, we have

$$< \bar{\partial}\frac{1}{Q}, \, \Delta\frac{G}{\tilde{H}}d\zeta > = \frac{1}{(2\pi i)^n} \int_{\Gamma_Q(\epsilon)} \frac{\Delta(\zeta)G(\zeta)d\zeta}{Q_1(\zeta_1)\ldots Q_n(\zeta_n)\tilde{H}(\zeta)} . \tag{4.45}$$

We compute the right-hand side of (4.45), one variable at a time. Let

$$H_1(\zeta) := \frac{\Delta(\zeta)G(\zeta)}{\tilde{H}(\zeta)} .$$

Fix $\zeta' = (\zeta_2, \ldots, \zeta_n)$, then

$$\frac{1}{(2\pi i)} \int_{|Q_1(\zeta_1)|=\epsilon_1} \frac{H_1(\zeta_1, \zeta')}{Q_1(\zeta_1)}d\zeta_1 =$$

$$= \sum_{k=1}^{s(1)} \left(\sum_{Q_{1,k}(\xi)-0} res_{\zeta_1=\xi}\left[\frac{H_1(\zeta_1, \zeta')/[Q_1(\zeta_1)/Q_{1,k}^{n_{1,k}}(\zeta_1)]}{(Q_{1,k}(\zeta_1))^{n_{1,k}}} \right] \right) . \tag{4.46}$$

Consider the kth term of the outside sum, and to simplify the notation we let $\nu = n_{1,k}$, $q = Q_{1,k}$, and A be the numerator of the corresponding quotient in the interior sum of (4.46). The zeroes of q are all simple, we denote them $\xi_1, \ldots, \xi_\mu$, and the function A is holomorphic at the ξ_j, since the different irreducible factors of Q_1 have no common zeroes. Let us factorize $q(t) = (t - \xi_1)R_1(t)$, R_1 is a polynomial in t with coefficients in $\mathbb{F}[\xi_1]$. Note that for a different root ξ_j, $q(t) = (t - \xi_j)R_j(t)$, where the polynomial $R_j \in \mathbb{F}[\xi_j][t]$ is obtained from R_1, just by replacing ξ_1 by ξ_j in the formulas defining the coefficients of R_1. The rational function $A(t)/q(t)$ has pole of order ν at $t = \xi_1$, so that

$$res_{t=\xi_1}\left[\frac{A(t)}{q(t)^\nu} \right] = \frac{1}{(\nu - 1)!} \frac{d^{\nu-1}}{dt^{\nu-1}}\left[\frac{A(t)}{R_1(t)^\nu} \right]_{t=\xi_1} . \tag{4.47}$$

It is clear that the right-hand side of (4.47) is a rational expression in ξ_1 and ζ', with coefficients in $\mathbb{F}$. Moreover, the corresponding value for ξ_j is obtained just by replacing ξ_1 by ξ_j, everywhere in the formulas. Therefore, by Galois theory,

$$\sum_{Q_{1,k}(\xi)=0} res_\xi\left(\frac{A(t)}{(Q_{1,k}(t))^{n_{1,k}}}\right) \tag{4.48}$$

is a rational function in $\mathbb{F}(\zeta')$.

The rational function $A \in \mathbb{F}(t, \zeta')$ has as denominator $\tilde{H}(t, \zeta')$, up to a polynomial factor that depends only on t. Hence, the residue in (4.47) has the form

$$\frac{B(\xi_1, \zeta')}{\tilde{H}(\xi_1, \zeta')^\nu}, \quad B \in \mathbb{F}(t)[\zeta'].$$

Therefore, a rational function in ζ', defined by (4.48), has no poles when ζ' is a zero of $Q_2(\zeta_2)\cdots Q_n(\zeta_n)$, so that we can repeat the process. One concludes that

$$< \bar{\partial}\frac{1}{P},\, Rd\zeta > \in \mathbb{F},$$

as we claimed.   □

In the algebraic setting, there are also interesting consequences of the integral formulas we gave in Chapter 3, §3, representing the complete sum of the residues of a holomorphic $\mathbb{C}^n$-valued map. In order to make some of these consequences explicit, we first place ourselves in the projective setting. The Grothendieck residue can also be introduced for functions holomorphic on an open subset of a complex manifold $\mathcal{X}$, defining a discrete variety. Let us consider the case when $\mathcal{X} = \mathbb{P}^n(\mathbb{C})$, suppose that $S_1, \ldots, S_n$ are homogeneous polynomials of $n+1$ variables, defining a zero-dimensional algebraic variety $\mathcal{V}$ in $\mathbb{P}^n(\mathbb{C})$, and suppose that ϕ is a meromorphic $(n, 0)$ form in $\mathbb{P}^n(\mathbb{C})$, whose poles lie in the projective hypersurface $\{S_1 \cdots S_n = 0\}$. Then, by Stokes's theorem, the complete sum of residues

$$< \bar{\partial}(1/S), \phi > = 0. \tag{4.49}$$

(see [7, p. 655–656].) This remark can be used to prove a very useful result of Jacobi [8], [9].

Proposition 4.12. *Suppose that $P_1, \ldots, P_n$ are polynomials of n variables such that the homogeneous polynomials $^hP_1, \ldots, ^h P_n$ define a discrete variety $\mathcal{V}$ in $\mathbb{P}^n(\mathbb{C})$, which lies entirely in $\mathbb{C}^n$. Let $Q \in \mathbb{C}[X_1, \ldots, X_n]$ be such that*

$$\delta := \deg(P_1) + \ldots + \deg(P_n) - \deg(Q) \geq n+1$$

Under these conditions,

$$< \bar{\partial}\frac{1}{P},\, Q\, d\zeta > = 0. \tag{4.50}$$

Proof. We return to the considerations introduced in Example 2, which followed Theorem 3.4. Consider the differential form in $\mathbb{C}^n$,

$$\omega = \frac{Q(\zeta)}{P_1(\zeta)\dots P_n(\zeta)}d\zeta.$$

Let us extend this differential form to $\mathbb{P}^n(\mathbb{C})$ by introducing homogeneous coordinates, $(\zeta_0', \dots, \zeta_n')$, with $\zeta_1 = \zeta_1'/\zeta_0', \dots, \zeta_n = \zeta_n'/\zeta_0'$,

$$\tilde{\omega} = \frac{(\zeta_0')^{\delta}\ {}^hQ(\zeta')}{{}^hP_1(\zeta')\dots {}^hP_n(\zeta')}\sum_{k=0}^{n}(-1)^k\frac{\zeta_k'd\zeta_{[k]}'}{(\zeta_0')^{n+1}}.$$

Since $\delta \geq n+1$, the differential form $\tilde{\omega}$ is a form on $\mathbb{P}^n(\mathbb{C})$ that is singular only on the projective hypersurface ${}^hP_1\cdots{}^hP_n = 0$. From (4.49) and the fact that the projective variety $\mathcal{V}$ is entirely contained in $\mathbb{C}^n$, and thus, coincides with $V = \{P_1 = \dots = P_n = 0\} \subset \mathbb{C}^n$, the statement (4.50) follows.   □

The hypotheses of the last proposition and, in particular, the fact that there should be no zeroes at infinity, are fairly restrictive. A weaker hyphothesis is to assume, for the $\mathbb{C}^n$-valued polynomial map P, that there are two constants $d, \gamma > 0$ such that, for all $\|\zeta\| \gg 0$,

$$\|P(\zeta)\| \geq \gamma\|\zeta\|^d \tag{4.51}$$

This condition implies the properness of the polynomial map

$$P : \mathbb{C}^n \longrightarrow \mathbb{C}^n.$$

It seems plausible that the property of being proper is actually equivalent to the inequality (4.51), and, probably, the best exponent d must be rational. Results of this type often depend on the Seidenberg-Tarski decision procedure ([13].) The following example shows that (4.51) is weaker than the hypotheses of Proposition 4.12. For $n = 2$, let

$$P_1(\zeta) = \zeta_1\zeta_2, \quad P_2(\zeta) = (\zeta_1 + 1)(\zeta_2 + 1).$$

It is easy to check that (4.51) holds with $d = 1$, while there are two points at ∞, since both polynomials have the same leading homogeneous term.

The following theorem will be very useful for us in Chapter 5.

Theorem 4.13. *Let $P_1, \dots, P_n$ be polynomials in $\mathbb{C}^n$ of degree at most D, satisfying the condition (4.51), then, for any polynomial Q and any multiindex β such that*

$$(|\beta| + n)d > \deg(Q) + (n-1)(D-d) + n, \tag{4.52}$$

we have

$$<\bar{\partial}\frac{1}{P^*(\beta+1)}, Q\,d\zeta >= 0. \tag{4.53}$$

Proof. Let us apply formula (3.33) for $\epsilon_j \sim N \gg 0$, $1 \le j \le n$, so that $\partial\Delta_\epsilon$ is connected and the variety V of common zeroes of the P_j lies inside Δ_ϵ. This is possible because the map P is proper. The integrand in (3.33) is

$$\frac{\bar{f}^{*\beta} \sum_{k=1}^n (-1)^{k-1} \bar{f}_k \overline{df_{[k]}} \wedge Q d\zeta}{\|f\|^{2(n+|\beta|)}},$$

which can be estimated by CN^q, for some positive constant $C = C(n, P, Q)$, with the exponent

$$q = \deg(Q) + (n-1)(D-1) - (|\beta| + 2n - 1)d.$$

Since, the surface volume of $\partial\Delta_\epsilon$ is essentially N^{2n-1}, the condition (4.52) is exactly what is needed to let $N \to \infty$ and obtain the vanishing of the residue. This concludes the proof. $\square$

To conclude this chapter, we would like to give a simple proof of a theorem of Macaulay [12],[14].

Theorem 4.14. *Let $S_1, \ldots, S_{n+1}$ be homogeneous polynomials in $\mathbb{C}^{n+1}$, having the origin as their only common zero. Let $\kappa = \deg(S_1) + \cdots + \deg(S_{n+1}) - n$, then*

$$X_0^\kappa \in I(S_1, \ldots, S_{n+1}).$$

Proof. The proof is based on the Cauchy-Weil formula. We consider sequences $\epsilon_j(k) \sim k^{-D_j}$, $1 \le j \le n+1$, where $D_j = \deg(S_j)$, defining Weil polyhedra $\Delta_\epsilon(k)$ corresponding to the S_j containing the origin. As in the second proof we gave of the Briançon-Skoda theorem (Remark 4.6), we have to show that

$$\lim_{k \to \infty} \int_{\Gamma_S(\epsilon(k))} \frac{\zeta_0^\kappa \, d\zeta}{S_1(\zeta) \cdots S_{n+1}(\zeta)} = 0.$$

We make the change of variables, $u = k\zeta$ in the last integral. Because of the homogeneity of the integrand, the integral becomes $O(1/k)$. This proves that the remainder term, when one divides ζ_0^κ using the Cauchy-Weil formula in any Weil polyhedron about the the origin, is always zero. Let us use (4.2) as a division formula in a fixed Weil polyhedron Δ about the origin, then one obtains

$$z_0^\kappa = \sum_{|k|>0} < \bar{\partial} \frac{1}{S(\zeta)^{*(k+\underline{1})}}, \zeta_0^\kappa \bigwedge_{j=1}^{n+1} g_j(z, \zeta) > S(z)^{*k}. \qquad (4.54)$$

The $g_{j,l}$, $1 \le j \le n+1$, $0 \le l \le n$, are the homogeneous Hefer polynomials of degree $D_j - 1$, in $2n+2$ variables, given by

$$g_{j,l}(z, \zeta) = \frac{S_j(\zeta_0, \ldots, \zeta_{l-1}, z_l, \ldots, z_n) - S_j(\zeta_0, \ldots, \zeta_l, z_{l+1}, \ldots, z_n)}{z_l - \zeta_l}. \qquad (4.55)$$

We remark that the S_j satisfy the conditions of Proposition 4.12, so that we can apply the same proposition to any of the polynomial maps $S^{k+\underline{1}}$, and see that the coefficient of $S(z)^{*k}$ is zero, whenever

$$k_1 D_1 + \cdots + k_{n+1} D_{n+1} \geq D_1 + \cdots + D_{n+1} - n + 1.$$

Recalling the notation $< k, D >= k_1 D_1 + \cdots + k_{n+1} D_{n+1}$, we see that (4.54) becomes the finite sum

$$z_0^{\kappa} = \sum_{\substack{|k|>0 \\ <k,D>\leq \kappa}} < \bar{\partial} \frac{1}{S(\zeta)^{*(k+\underline{1})}}, \zeta_0^{\kappa} \bigwedge_{j=1}^{n+1} g_j(z,\zeta) > S(z)^{*k}. \qquad (4.56)$$

Since $|k| > 0$, this is an explicit division formula in the ideal $I(S_1, \ldots, S_{n+1})$. This ends the proof. $\qquad\qquad\qquad\qquad\qquad\qquad\qquad\qquad\qquad\qquad\qquad\Box$

References for Chapter 4

[1] L.A. Aizenberg and A.P. Yuzhakov, *Integral Representation in Multidimensional Complex Analysis*, Transl. Amer. Math. Soc. 58, 1980.

[2] L.A. Aizenberg, The multidimensional logarithmic residue and its applications, *Modern Problems of Mathematics (Fundamental Directions)*, vol. 8, Moscow 1985 (in Russian); English translation in *Encyclopedia of Math. Sci.*, vol. 8 (*Several Complex Variables II*), Springer-Verlag.

[3] C.A. Berenstein, R. Gay, and A. Yger, Analytic continuation of currents and division problems, Forum Math.1(1989), 15–51.

[4] C.A. Berenstein and B.A. Taylor, Interpolation problems in $\mathbb{C}^n$ with applications to harmonic analysis, J. Analyse Math. 38(1980) 188–254.

[5] C.A. Berenstein and A. Yger, Effective Bezout identities in $\mathbb{Q}[z_1, \ldots, z_n]$, Acta Mathematica, 166(1991), 69–120.

[6] C.A. Berenstein and A. Yger, Une formule de Jacobi et ses conséquences, Ann. Sci. Ec. Norm. Sup. Paris 24(1991), 363–377.

[7] P. Griffiths and J. Harris, *Principles of Algebraic Geometry*, Wiley-Interscience, New York, 1978.

[8] C.G.J. Jacobi, Theoremata nova algebraica circa systema duarum aequationum inter duas variabiles propositarum, Gesammelte Werke, Band III, 285–294.

[9] C.G.J. Jacobi, De relationibus, quae locum habere debent inter puncta intersectionis duarum curvarum vel trium superficierum algebraicarum dati ordinis, simul cum enodatione paradoxi algebraici, Gesammelte Werke, Band III, 329–354.

[10] J. Kollár, Sharp effective Nullstellensatz, Journal of Amer. Math. Soc. 1(1988), 963–975.

[11] J. Lipman and B. Teissier, Pseudorational local rings and a theorem of Briançon-Skoda about integral closures of ideals, Michigan Math. J. 28(1981), 97 115.

[12] F.S. Macaulay, *The algebraic theory of modular systems*, Cambridge Univ. Press, Cambridge, 1916.

[13] A. Seidenberg, Constructions in Algebra, Trans. Amer. Math. Soc., 197(1974), 273–313.

[14] B. Shiffman, Degree bounds for division problem in polynomial ideals, Michigan Math. J. 36(1989), 163–171.

[15] A.K. Tsikh, *Multidimensional Residues and Their Applications*, Transl. Amer. Math. Soc. 103, 1992.

[16] B.L. van der Warden, *Modern Algebra*, Springer-Verlag, New York, 1979.

[17] V.S. Vladimirov, *Methods of the Theory of Functions of several Complex Variables*, M.I.T. Press, 1966.

[18] A. Weil, L'intégrale de Cauchy et les fonctions de plusieurs variables, Math. Ann. 111(1935), 178–182.

[19] A.P. Yuzhakov, On the appllication of the total sum of residues with respect to a polynomial mapping in $\mathbb{C}^n$, Dokl. Akad. Nauk SSSR 275(1984), 817–820; English transl. in Soviet Math. Dokl. 29(1984).

[20] A.P. Yuzhakov, Methods of calculating multidimensional residues, *Modern Problems of Mathematics (Fundamental Directions)*, vol. 8, Moscow 1985 (in Russian); English translation in *Encyclopedia of Math. Sci.*, vol. 8 (*Several Complex Variables II*), Springer-Verlag, New York.

Chapter 5
Applications to Commutative Algebra and Harmonic Analysis

§1 An analytic version of the algebraic Nullstellensatz

Let $\mathbb{F}$ be a subfield of $\mathbb{C}$, and $P_1, \ldots, P_m$ be elements of $\mathbb{F}[X_1, \ldots, X_n]$. If $Q \in \mathbb{F}[X_1, \ldots, X_n]$ vanishes on $V = \{P_1 = \ldots = P_m = 0\}$, some power of Q lies in the ideal I generated by $P_1, \ldots, P_m$. This classical result, which is known as the global algebraic Nullstellensatz, can be reduced to the following special case, where the polynomials have no common zeroes in $\mathbb{C}^n$. In that case, solving the Nullstellensatz corresponds to the problem of finding m polynomials $A_1, \ldots, A_m$ in $\mathbb{F}[X_1, \ldots, X_n]$, such that

$$ 1 = A_1 P_1 \quad \ldots + A_m P_m . \tag{5.1} $$

This polynomial equation (5.1) is called the (algebraic) Bezout equation.

The classical method to solve this equation for $n = 1$ is to use the division algorithm of Euclides. As we have seen in Chapter 1, §5, one can propose completely different methods, which are not based on a recurrence algorithm, as the Euclidean algorithm, but on direct explicit formulas. Note that the formula we gave in Example 1 was completely analytic, while the variant we proposed in Example 2, since it involved residue currents, seemed to be more algebraic in nature. (At least, the identity (5.1) could be solved in an algebraic closure of $\mathbb{F}$.) On the other hand, these last two methods could be significantly more complex to carry on than the Euclidean division algorithm. (In fact, in the one variable case this algorithm is optimal from the point of view of complexity theory.)

One needs to look at the multidimensional problem to realize how far theoretical results can be from implementation. In this case also, there is an old algorithm to solve the Bezout identity, using elimination theory (see [74]). The explicit construction of $A_1, \ldots, A_m$, done this way, involves computations of resultants, and one can measure the complexity of the problem by looking at the theoretical bounds one can expect on the $\deg(A_j)$ (in terms of $D = \max \deg(P_k)$); such bounds have been computed explicitly, because of applications to Transcendental Number Theory, and are

$$ \deg(A_j) \leq 2(2D)^{2^{n-1}} . \tag{5.2} $$

(see, e.g.,[51,§4].) Of course, there are particular situations where one has much better a priori bounds. Such is the case, for instance, when the homogeneous polynomials $^h P_j$, introduced in the last chapter, define an empty variety in the projective space. The theorem of Macaulay (Theorem 4.14) [50], which we proved when $m = n + 1$ (to which one can always reduce the question), shows that one can find A_j solving (5.1) and

$$\deg(A_j P_j) \leq D_1 + \cdots + D_{n+1} - n.$$

What is interesting in this special case, is that we gave also an explicit formula (just set $z_0 = 1$ in (4.56)), involving multidimensional residues, and solving the problem with essentially the same kind of bounds. Given that there is a large gap between the Macaulay situation and the bounds (5.2), it has been a longstanding question to find out which were the sharp bounds for these degrees, and also what should be complexity of solving the Bezout equation. With respect to the bounds of the degrees, the following example of Masser-Philippon (see [23], [48], for background and references),

$$P_1(z) = z_1^D, \ P_2(z) = z_1 - z_2^D, \ldots, P_{n-1}(z) = z_{n-2} - z_{n-1}^D, \ P_n(z) = 1 - z_{n-1} z_n^{D-1}$$

shows, by restricting the identity (5.1) to a convenient algebraic curve, that

$$\max \deg(A_j) \geq D^n - D^{n-1}.$$

Thus, there is no hope to obtain bounds better than D^n (in general, the product of the n largest degrees), which corresponds roughly to the Bezout estimates found in Theorem 2.22.

Solving the Bezout equation is a particular case of a more general problem, the membership problem, that will be discussed in Section 2 below. In the membership problem, one has to find a way to check whether a polynomial Q belongs to the ideal I generated by $P_1, \ldots, P_m$; furthermore, one wants to obtain explicit solutions A_j (sometimes called quotients) of the equation

$$Q = \sum_{j=1}^{m} A_j P_j, \tag{5.3}$$

if they exist. The membership problem (5.3) can be solved immediately, if one has a Gröbner basis for the ideal I [26]. An algorithm to construct Gröbner bases, which generalizes to $n > 1$ the Euclidean division algorithm and has been widely implemented, is the Buchberger algorithm. On the downside, even for the solution of Bezout identity, there are no a priori bounds on the degrees of the solutions A_j of (5.1) obtained by the Buchberger algorithm.

Let us return to the question of what are the sharp bounds for the $\deg(A_j)$ in (5.1). Using transcendental methods developped by Nesterenko [54], [55] based on the theory of Chow's forms, Brownawell gave in 1987 [23] almost optimal bounds for this problem. In order to state his theorem, we need to remind the reader of the concept of a regular sequence in an integral domain

$\mathcal{R}$. It is an ordered sequence $a_1, \ldots, a_k$ in $\mathcal{R}$ such that a_{j+1} is not a zero divisor in the quotient $\mathcal{R}/I(a_1, \ldots, a_j)$, for $0 \leq j \leq k - 1$.

Theorem 5.1. *Let $P_1, \ldots, P_m$ be m polynomials in $\mathbb{F}[X_1, \ldots, X_n]$, defining a regular sequence, without common zeroes, and ordered so that*

$$D_1 := \deg(P_1) \geq \ldots \geq D_m := \deg(P_m).$$

Then, there is constant $c > 0$ such that for all $\zeta \in \mathbb{C}^n$

$$\|P(\zeta)\| \geq c(1 + \|\zeta\|)^{1-(n-1)D_1 \cdots D_\mu} \quad \mu = \min(n, m). \tag{5.4}$$

Moreover, for any finite family $\mathcal{P} \subset \mathbb{F}[X_1, \ldots, X_n]$, let

$$V = \{\zeta \in \mathbb{C}^n : P(\zeta) = 0, \, \forall P \in \mathcal{P}\}.$$

If $d(\zeta, V)$ is the Euclidean distance function, we denote

$$\tilde{d}(\zeta, V) = \min\{1, d(\zeta, V)\}, \quad m = \#(\mathcal{P}), \quad \mu = \inf(n, m).$$

Brownawell [24] extended the previous theorem into a global version of the Lojasiewicz inequalities (see Chapter 3, §2.)

Theorem 5.2. *For any finite family $\mathcal{P}$ of polynomials, let $D := \max_\mathcal{P} \deg(P)$; there is a positive constant C such that*

$$\left(\frac{\tilde{d}(\zeta, V)}{1 + \|\zeta\|^2}\right)^{(n+1)^2 D^\mu} \leq C \max_\mathcal{P} |P(\zeta)| \quad \forall \zeta \in \mathbb{C}^n. \tag{5.5}$$

Remark 5.3. For a regular sequence as in Theorem 5.1, Brownawell also proved in [23] that there is $c > 0$ such that

$$\log \max_{1 \leq j \leq m}\left(\frac{|P_j(\zeta)|}{(1 + \|\zeta\|)^{D_j}}\right) \geq -c - nD_1 \ldots D_\mu \log\left(\frac{1 + \|\zeta\|}{\tilde{d}(\zeta, V)}\right). \tag{5.6}$$

Using the results of Skoda [70] about L^2 estimates for division problems (see [8] for a survey of these methods,) Brownawell used Theorem 5.1 to obtain reasonably sharp bounds for the degrees of some solutions of the Bezout equation (5.1), namely,

$$\deg(A_j) < n\mu D_1 \cdots D_\mu + \mu D. \tag{5.7}$$

Later, Kollár [48], Heintz and collaborators [27], [28], [29], Shiffman [68], Ji-Kollár-Shiffman [46], and others, studied the same problems and sharpened the bounds (5.4)–(5.7), replacing the analytic arguments of Brownawell by ideas from commutative algebra, algebraic geometry, etc.

The disadvantage of using Skoda's results is that it is only an existence theorem, which ignores the underlying field $\mathbb{F}$ and the algebraic structure of the problem. Such was also the case in Example 1 of Chapter 1, §5, but not in Example 2. What we propose here is to give a solution to the Bezout equation, inspired by the latter example, so that we can keep track of the field. We prove first the following proposition.

Proposition 5.4. *Let* $P_1, \ldots, P_{n+1}$ *be polynomials without common zeroes in* $\mathbb{C}^n$, *of degrees at most* D, *and with the property that there are strictly positive constants* γ, d, *so that for all* ζ, $\|\zeta\| \gg 0$,

$$\max_{1 \le j \le n} |P_j(\zeta)| \ge \gamma \|\zeta\|^d . \tag{5.8}$$

Let $g_{j,k}$ *denote the Hefer polynomials defined by (4.55) and let* Δ *be the determinant of order* $(n+1) \times (n+1)$

$$\Delta(z, \zeta) = \begin{vmatrix} g_{1,1}(z,\zeta) & \cdots & g_{n,1}(z,\zeta) & g_{n+1,1}(z,\zeta) \\ \cdots & \cdots & \cdots & \cdots \\ g_{1,n}(z,\zeta) & \cdots & g_{n,n}(z,\zeta) & g_{n+1,n}(z,\zeta) \\ P_1(z) - P_1(\zeta) & \cdots & P_n(z) - P_n(\zeta) & P_{n+1}(z) \end{vmatrix}$$

Then, for any $z \in \mathbb{C}^n$ *and any integer* q *such that*

$$d(q+1) \ge 2(n-1)(D-d) + 1 , \tag{5.9}$$

one has

$$1 = \sum_{|k| \le q} < (\bar{\partial} \frac{1}{P_1^{k_1+1}} \wedge \cdots \wedge \bar{\partial} \frac{1}{P_n^{k_n+1}})(\zeta), \frac{\Delta(z,\zeta)}{P_{n+1}(\zeta)} d\zeta > (P_1^{k_1} \cdots P_n^{k_n})(z) . \tag{5.10}$$

Remark 5.5. The formula (5.10) is a Bezout identity because, developping Δ by the last row, one sees that

$$\Delta(z, \zeta) = \sum_{j=1}^{n} u_j(z, \zeta) P_j(\zeta) + \sum_{l=1}^{n+1} v_l(z, \zeta) P_l(z) , \tag{5.11}$$

and the action of the residue current corresponding to $k = \underline{0}$ kills the first sum in (5.11).

Proof. We apply Proposition 4.4, with $f_j = P_j$ for $1 \le j \le n$ and we choose the auxiliary function τ as

$$\tau(z, \zeta) = \varphi(\zeta) \frac{\bar{g}_{n+1}(z, \zeta)}{P_{n+1}(\zeta)} ,$$

where φ is a test function, identically one on the finite set of common zeroes of $P_1, \ldots, P_n$, and so that P_{n+1} does not vanish on $\mathrm{supp}(\varphi)$. The differential form Θ, associated to τ and to the Hefer polynomials $g_{j,k}$, $1 \le j,k \le n$, as in Proposition 4.4, is given in a neighborhood of the common zeroes of the first n polynomials by

$$\Theta(z,\zeta) = \frac{\Delta(z,\zeta)d\zeta}{P_{n+1}(\zeta)}.$$

We write formula (4.14) for large N and $U = B(0,R)$, with R large. Since τ has compact support, we obtain

$$1 = \sum_{|k| \le N-n} < (\bar{\partial}\frac{1}{P_1^{k_1+1}} \wedge \ldots \wedge \bar{\partial}\frac{1}{P_n^{k_n+1}})(\zeta), \frac{\Delta(z,\zeta)}{P_{n+1}(\zeta)}d\zeta > (P_1^{k_1} \ldots P_n^{k_n})(z)$$

$$+ \frac{1}{(2\pi i)^n} \int_{\|\zeta\|=R} \sum_{\alpha_0+\alpha_1=n-1} \binom{N}{\alpha_1} \left(\frac{\sum_{j=1}^n \bar{P}_j P_j(z)}{\sum_{j=1}^n |P_j|^2} \right)^{N-\alpha_1} \frac{S \wedge \bar{\partial}S^{\alpha_0} \wedge \bar{\partial}A^{\alpha_1}}{\|z-\zeta\|^{2(\alpha_0+1)}}.$$

$$(5.12)$$

Recall that here

$$A = \frac{\sum_{j=1}^n \bar{P}_j g_j(z,\zeta)}{\sum_{j=1}^n |P_j|^2}.$$

For z fixed, we can estimate on the sphere $\|\zeta\| = R$ the term

$$\left| \frac{\sum_{j=1}^n \bar{P}_j(\zeta)P_j(z)}{\sum_{j=1}^n |P_j(\zeta)|^2} \right| \le \frac{C(z)}{R^d},$$

and

$$\|\bar{\partial}A(z,\zeta)\| \le C(z) R^{2(D-d-1)},$$

with $C(z)$, a positive constant depending on z, uniformly bounded on compacts. These estimates imply that the integral in (5.12) tends to zero, when z is fixed and $R \to \infty$, as soon as $Nd \ge (n-1)(2D-d)+1$. This condition is fulfilled by $N = q + n$, whenever the inequality (5.9) is satisfied by q. This concludes the proof of the proposition.   □

Remark 5.6. The explicit division formula (5.10) has the advantage that if the polynomials P_j have coefficients in the subfield $\mathbb{F}$ of $\mathbb{C}$, then, from Proposition 4.11, the quotients A_j have coefficients in the same field. It follows that the statement of this proposition holds for any field of characteristic zero.

The same method, i.e., an argument based on the use of Proposition 4.4, can be used to solve division problems in spaces of entire functions with growth conditions, as in Example 3, Chapter 1, §5, see also [10] and [69], where they were implemented in a Connection Machine. In the algebraic case, there is a simpler way to obtain (5.10) with slightly worse conditions on q. This can be done using the Cauchy-Weil formula. As in the proof of Theorem 4.13, let us apply (4.2) to the function $h \equiv 1$, assuming that $\epsilon_j \sim N \gg 0$ (i.e., the set

$\Delta_{P_1,\ldots,P_n}(\epsilon) = \{|P_j| < \epsilon_j \; 1 \leq j \leq n\}$ is connected and contains all the common zeroes of $P_1,\ldots,P_n$.) For any z in $\Delta_{P_1,\ldots,P_n}(\epsilon)$, one obtains

$$1 = \sum_{k \in \mathbb{N}^n} < (\bar{\partial}\frac{1}{P_1^{k_1+1}} \wedge \ldots \wedge \bar{\partial}\frac{1}{P_n^{k_n+1}})(\zeta), \bigwedge_{j=1}^{n} g_j(z,\zeta) > (P_1^{k_1} \ldots P_n^{k_n})(z).$$

$$(5.13)$$

We now use theorem 4.13 and claim that, whenever $|k|d > (2n-1)(D-d)$, the coefficient of $(P_1^{k_1} \ldots P_n^{k_n})(z)$ in the development (5.13) is equal to zero; therefore, one has the following polynomial identity

$$1 = \sum_{\substack{k \in \mathbb{N}^n \\ |k|d \leq (2n-1)(D-d)}} < (\bar{\partial}\frac{1}{P^{k+\underline{1}}})(\zeta), \bigwedge_{j=1}^{n} g_j(z,\zeta) > P^k(z), \qquad (5.14)$$

where only the polynomials $P_1,\ldots,P_n$ appear. Let us make the following transformation on the determinant Δ; if $\mathcal{L}_1, \ldots \mathcal{L}_n$ denote the first n rows of this determinant, just substract from the last row the linear combination

$$(z_1 - \zeta_1)\mathcal{L}_1 + \cdots + (z_n - \zeta_n)\mathcal{L}_n.$$

One obtains in this way

$$\Delta(z,\zeta)d\zeta = P_{n+1}(\zeta) \bigwedge_{j=1}^{n} g_j(z,\zeta).$$

Observe that, for any multiindex k, one can rewrite the coefficient of the polynomial $(P_1^{k_1} \ldots P_n^{k_n})(z)$ in the expansion (5.14) as

$$< (\bar{\partial}\frac{1}{P_1^{k_1+1}} \wedge \ldots \wedge \bar{\partial}\frac{1}{P_n^{k_n+1}})(\zeta), \frac{P_{n+1}(\zeta) \bigwedge_{j=1}^{n} g_j(z,\zeta)}{P_{n+1}(\zeta)} >, \qquad (5.15)$$

(since P_{n+1} does not vanish in some neighborhood of the support of the residue current involved in (5.15)); one obtains then the following Bezout identity

$$1 = \sum_{\substack{k \in \mathbb{N}^n \\ |k|d \leq (2n-1)(D-d)}} < (\bar{\partial}\frac{1}{P^{k+\underline{1}}})(\zeta), \frac{\Delta(z,\zeta)d\zeta}{P_{n+1}(\zeta)} > P^k(z) \qquad (5.16)$$

which is just another version of (5.10). Note that the sum here involves more terms than (5.10) since

$$2(n-1)D - (2n-1)d + 1 < (2n-1)(D-d),$$

when $D > 1$. We obtain therefore another approach to our Bezout formula; unfortunately, this approach fails (because one has to interchange limits) when the set of common zeroes of $P_1, \ldots, P_n$ is not finite, that is when one deals with transcendental functions instead of polynomials.

The condition (5.8) in Proposition 5.4 is not so restrictive as it may seem, due to the following consequence of E. Noether's Normalization Theorem (combined with the Lojasiewicz-Brownawell inequality (5.6).)

Proposition 5.7. *Let $p_1, \ldots, p_n$ be n polynomials in $\mathbb{C}[X_1, \ldots, X_n]$, which define a system in normal position, i.e., for any subset J of $\{1, \ldots, n\}$, if $W_J = \{p_j = 0, j \in J\}$ then $\mathrm{codim}(W_J) = \#(J)$, and let $\delta = \prod_{j=1}^{n} \deg(p_j)$. There exist n linear forms $L_1, \ldots, L_n$ with integral coefficients so that for any $N \in \mathbb{N}^*$, there is a strictly positive constant γ_N with the property that, if $\|z\| \gg 0$,*

$$\max_{1 \leq j \leq n} |L_j(z)^{nN\delta} p_j(z)| \geq \gamma_N \|z\|^{n(N-1)\delta} . \tag{5.17}$$

Proof. We give here a sketch of the proof from [12, Lemma 5.3]. Recall the well-known fact [40] that given an algebraic variety W of codimension $l < n$, after a generic linear change of coordinates, there are constants $C, K > 0$ such that, for any $w \in W$ with $\|w\| \geq C$, one has

$$|w_1| + \ldots + |w_l| \leq K(|w_{l+1}| + \cdots + |w_n|) . \tag{5.18}$$

This is just a consequence of Noether's Normalization Theorem [76], and one can show that the inverse of the matrix of this change of coordinates can be chosen to have integral coefficients. Furthermore, given a finite family of algebraic varieties, one can use the same matrix, and the same constants, so that any variety W of this family is contained in a cone of the form (5.18), with $l = \mathrm{codim}(W)$. Thus, we can find an invertible matrix A, with integral coefficients, and constants $C, K > 0$ such that, if $\|w\| \geq C$, $l = \#(J) < n$ and $Aw \in W_J$, then (5.18) holds. Note that for any $J \subset \{1, \ldots, n\}$, the polynomials $p_j \in J$ form a regular sequence, independently of the ordering. So, we can apply the Lojasiewicz-Brownawell inequality (5.6) to any such sequence of polynomials. Therefore, there exists $\epsilon > 0$ such that, for any index set J, $\#(J) = l < n$, and any w satisfying

$$\|w\| \geq C + 1 \quad \text{and} \quad \max_{j \in J} |p_j(Aw)| \leq \epsilon(1 + \|w\|)^{-n\delta} ,$$

one has

$$|w_1| + \ldots + |w_l| \leq (K + 1)(|w_{l+1}| + \ldots + |w_n|) . \tag{5.19}$$

Moreover, from the inequality (5.6) applied to the whole sequence $p_1, \ldots, p_n$, we have, up to some possible modification of the values C, K, ϵ, that for any w with $\|w\| \geq C$,

$$\max_{1 \leq j \leq n} |p_j(Aw)| \geq \epsilon(1 + \|w\|)^{-n\delta} . \tag{5.20}$$

Let B a $n \times n$ matrix with integer coefficients such that none of its minors vanishes, and let us denote by ρ the maximum modulus of all the minors of B.

Let M be an integer such that $M > (K + 1)n\rho$. We consider the linear forms Λ_j, $j = 1, \ldots, n$ defined by

$$\Lambda_j(w) = b_{j,1}w_1 + Mb_{j,2}w_2 + \cdots + M^{n-1}b_{j,n}w_n \,,$$

where $b_{j,k}$ are coefficients of B. It follows from Cramer's rule that there is a constant $\epsilon_0 > 0$ such that for any $k < n$, the inequality

$$|w_1| + \ldots + |w_k| \leq (K + 1)(|w_{k+1}| + \ldots + |w_n|) \tag{5.21}$$

implies

$$\min_{J: \#(J)=n-k} \left(\sum_{j \in J} |\Lambda_j(w)| \right) \geq \epsilon_0 \|w\| \,. \tag{5.22}$$

We now define

$$L_j(z) = \Lambda_j(\det(A)A^{-1}z) \,.$$

Let us show that these linear forms satisfy the requirements of the proposition. Let z be such that $\|A^{-1}z\| \geq C$, then, (5.20) implies that there is an index set $J = J(z)$ with $\#(J) = l(z) < n$ such that

$$\max_{j \in J} |p_j(z)| < \epsilon(1 + \|A^{-1}z\|)^{-n\delta} \,. \tag{5.23}$$

and

$$\min_{j \notin J} |p_j(z)| \geq \epsilon(1 + \|A^{-1}z\|)^{-n\delta} \,. \tag{5.24}$$

It follows from (5.19), (5.22), and (5.24) that for any $N \in \mathbb{N}^*$, one has for the z we are considering, that

$$\sum_{j \notin J(z)} |L_j(z)|^{Nn\delta} |p_j(z)| \geq \kappa(\epsilon_0 |\det(A)|)^{Nn\delta} \epsilon \|A^{-1}z\|^{(N-1)n\delta} \,,$$

where $\kappa > 0$ is an absolute constant depending only on n. This concludes the proof of the Proposition 5.7.   □

We would like now to obtain the same result for a system of n polynomials defining a discrete variety in $\mathbb{C}^n$. We have the following proposition, due to Elkadi [38].

Proposition 5.8. *Let $q_1, \ldots, q_n$ be polynomials defining a complete intersection in $\mathbb{C}^n$. Let $\tilde{\delta} = \prod\limits_{j=1}^{n} \deg(q_j)$. There are n linear combinations of them, $p_1, \ldots, p_n$, with integral coefficients, and n linear forms, $L_1, \ldots, L_n$, also with integral coefficients, so that for any $N \in \mathbb{N}^*$, there is a strictly positive constant γ_N with the property that, if $\|z\| \gg 0$,*

$$\max_{1 \leq j \leq n} |L_j(z)^{nN\tilde{\delta}} p_j(z)| \geq \gamma_N \|z\|^{n(N-1)\tilde{\delta}} \,. \tag{5.25}$$

Furthermore, the q_j can also be expressed as linear combinations of the p_j.

Proof. We can assume that the sequence $\deg(q_j)$ is decreasing. Then, using the pigeonhole principle, as in [51,§4], one can find a triangular, invertible matrix R, with integral coefficients, such that the polynomials $s_1, \ldots, s_n$ defined by $R\vec{q} = \vec{s}$ form a regular sequence. Note that $\deg(s_j) = \deg(q_j)$. To simplify the notation, we will call this new sequence again q_j.

The construction of n linear combinations of the q_j defining a system p_j in normal position is also based on the pigeonhole principle; we refer to [12, Lemma 5.2, p.102], [51]. This construction can be made with the additional property: let $[\alpha_{j,k}]$ be the $n \times n$ matrix of coefficients of the linear combinations, we can assume that none of the minors of this matrix is zero. Unfortunately, at this point, all we can say about the degrees of the polynomials p_j is that they are bounded by the maximum of $D_j := \deg(q_j)$, which is not good enough to obtain the estimate (5.25), just using the previous proposition.

Let $J \subset \{1, ..., n\}, 1 \le k := \#(J) \le n-1$. Given collection of polynomials $\{p_j\}_{j \in J}$, we claim we can find a system of polynomials $\{\tilde{p}_{J,j}\}_{j \in J}$, such that for all $j \in J$, $\deg(\tilde{p}_{J,j}) \le D_j$ and all the $\tilde{p}_{J,j}$ are linear combinations of $p_j, j \in J$, and conversely. Let us assume, for example, that $J = \{1, ..., k\}$, and omit the index J. Recall that we assumed that the D_j were in decreasing order. We start with $\tilde{p}_1 = p_1$. In order to construct $\tilde{p}_2$, we choose a linear combination of p_1 and p_2, which is free of q_1, that is

$$\tilde{p}_2 = \alpha_{1,1}p_2 - \alpha_{2,1}p_1 = \begin{vmatrix} \alpha_{1,1} & \alpha_{1,2} \\ \alpha_{2,1} & \alpha_{2,2} \end{vmatrix} q_2 + \cdots + \begin{vmatrix} \alpha_{1,1} & \alpha_{1,n} \\ \alpha_{2,1} & \alpha_{2,n} \end{vmatrix} q_n.$$

In order to obtain $\tilde{p}_3$, one constructs a combination of p_3 and p_1 which is free of q_1, namely,

$$p_2^{(3)} = \alpha_{1,1}p_3 - \alpha_{3,1}p_1 = \begin{vmatrix} \alpha_{1,1} & \alpha_{1,2} \\ \alpha_{3,1} & \alpha_{3,2} \end{vmatrix} q_2 + \cdots + \begin{vmatrix} \alpha_{1,1} & \alpha_{1,n} \\ \alpha_{3,1} & \alpha_{3,n} \end{vmatrix} q_n$$

and then

$$\tilde{p}_3 = \begin{vmatrix} \alpha_{1,1} & \alpha_{1,2} \\ \alpha_{2,1} & \alpha_{2,2} \end{vmatrix} p_2^{(3)} - \begin{vmatrix} \alpha_{1,1} & \alpha_{1,2} \\ \alpha_{3,1} & \alpha_{3,2} \end{vmatrix} \tilde{p}_2$$

$$= \alpha_{1,1} \left(\begin{vmatrix} \alpha_{1,1} & \alpha_{1,2} & \alpha_{1,3} \\ \alpha_{2,1} & \alpha_{2,2} & \alpha_{2,3} \\ \alpha_{3,1} & \alpha_{3,2} & \alpha_{3,3} \end{vmatrix} q_3 + \cdots + \begin{vmatrix} \alpha_{1,1} & \alpha_{1,2} & \alpha_{1,n} \\ \alpha_{2,1} & \alpha_{2,2} & \alpha_{2,n} \\ \alpha_{3,1} & \alpha_{3,2} & \alpha_{3,n} \end{vmatrix} q_n \right)$$

This procedure allows us to construct $\tilde{p}_1, \ldots, \tilde{p}_k$ one by one. An inductive argument shows that, for $1 \le l \le k$,

$$\tilde{p}_l = p_1 p_2 ... p_{l-2} (\rho_l q_l + \rho_{l,l+1} q_{l+1} + ... + \rho_{l,n} q_n),$$

where

$$
\rho_j = \begin{vmatrix} \alpha_{1,1} & \alpha_{1,2} & \cdots & \alpha_{1,j} \\ \alpha_{2,1} & \alpha_{2,2} & \cdots & \alpha_{2,j} \\ \vdots & \vdots & \vdots & \vdots \\ \alpha_{j,1} & \alpha_{j,2} & \cdots & \alpha_{j,j} \end{vmatrix} \qquad \rho_{l,j} = \begin{vmatrix} \alpha_{1,1} & \cdots & \alpha_{1,l-1} & \alpha_{1,j} \\ \alpha_{2,1} & \cdots & \alpha_{2,l-1} & \alpha_{2,j} \\ \vdots & \vdots & \vdots & \vdots \\ \alpha_{l,1} & \cdots & \alpha_{l,l-1} & \alpha_{l,j} \end{vmatrix}
$$

This follows from the following identity

$$
\begin{vmatrix} \alpha_{1,1} & \cdots & \alpha_{1,l-1} & \alpha_{1,l} \\ \vdots & \vdots & \vdots & \vdots \\ \alpha_{l-1,1} & \cdots & \alpha_{l-1,l-1} & \alpha_{l-1,l} \\ \alpha_{l,1} & \cdots & \alpha_{l,l-1} & \alpha_{l,l} \end{vmatrix} \begin{vmatrix} \alpha_{1,1} & \cdots & \alpha_{1,l-1} & \alpha_{1,r} \\ \vdots & \vdots & \vdots & \vdots \\ \alpha_{l-1,1} & \cdots & \alpha_{l-1,l-1} & \alpha_{l-1,r} \\ \alpha_{l+1,1} & \cdots & \alpha_{l+1,l-1} & \alpha_{l+1,r} \end{vmatrix} -
$$

$$
- \begin{vmatrix} \alpha_{1,1} & \cdots & \alpha_{1,l-1} & \alpha_{1,l} \\ \vdots & \vdots & \vdots & \vdots \\ \alpha_{l-1,1} & \cdots & \alpha_{l-1,l-1} & \alpha_{l-1,l} \\ \alpha_{l+1,1} & \cdots & \alpha_{l+1,l-1} & \alpha_{l+1,l} \end{vmatrix} \begin{vmatrix} \alpha_{1,1} & \cdots & \alpha_{1,l-1} & \alpha_{1,r} \\ \vdots & \vdots & \vdots & \vdots \\ \alpha_{l-1,1} & \cdots & \alpha_{l-1,l-1} & \alpha_{l-1,r} \\ \alpha_{l,1} & \cdots & \alpha_{l,l-1} & \alpha_{l,r} \end{vmatrix} =
$$

$$
= \begin{vmatrix} \alpha_{1,1} & \cdots & \alpha_{1,l-1} \\ \vdots & \vdots & \vdots \\ \alpha_{l-1,1} & \cdots & \alpha_{l-1,l-1} \end{vmatrix} \begin{vmatrix} \alpha_{1,1} & \cdots & \alpha_{1,l} & \alpha_{1,r} \\ \alpha_{2,1} & \cdots & \alpha_{2,l} & \alpha_{2r} \\ \vdots & \vdots & \vdots & \vdots \\ \alpha_{l+1,1} & \cdots & \alpha_{l+1,l} & \alpha_{l+1,r} \end{vmatrix}
$$

where the determinants in the right-hand side are of order l, and those in the left hand-side are of respective orders $l - 1$ and $l + 1$. Since all minors of the matrix $[\alpha_{j,k}]$ are non zero, the new polynomials $\tilde{p}_j$ are linear combinations of the p_j and conversely. In particular, there is a positive constant c (depending only on the minors involved in the definition of the collection of all $\tilde{p}_{J,j}$, for all J and $j \in J$) such that for any z one has

$$
\max_{j \in J} |\tilde{p}_{J,j}(z)| \le c \max_{j \in J} |p_j(z)|. \tag{5.26}
$$

Let us now consider all such families $(\tilde{p}_{J,j})_{j \in J}$ for all possible $J \subset \{1, ..., n\}$, $\#(J) \le n - 1$. As in the proof of Proposition 5.7, we construct a matrix A with integer coefficients and three constants $\epsilon, C, K > 0$ such that, for every J, $l = \#(J)$, the set

$$
\mathcal{X}_J^{(\epsilon)} := \left\{ \|w\| > C : \max_{j \in J} |\tilde{p}_{J,j}(Aw)| < \frac{\epsilon}{(1 + \|w\|)^{n\bar\delta}} \right\}
$$

is included in the cone

$$
\mathcal{Y}_l := \{w : |w_1| + ... + |w_l| \le K(|w_{l+1}| + ... + |w_n|)\}.
$$

As before, this follows from (5.6). We now define the linear forms Λ_j, adapted to our new family of cones $\mathcal{Y}_l$, as in the proof of the preceding proposition.

Since the p_j are linear combinations of the q_j, and conversely, it is also a consequence of (5.6) (applied to the regular sequence q_j), that after a suitable increase of the constants C, K, and decrease of ϵ, one has for $\|w\| > C$,

$$\max_{1 \le j \le n} |p_j(Aw)| \ge \frac{\epsilon}{(1 + \|w\|)^{n\bar\delta}} .$$

Hence, for $\eta \ll \epsilon$, the set $\{\|w\| > C\}$ can be written as the disjoint union of the sets

$$\mathcal{Z}_J := \left\{ \|w\| > C : |p_j(Aw)| < \frac{\eta}{(1 + \|w\|)^{n\bar\delta}} \quad \text{if} \quad j \in J \right.$$

$$\left. \text{and} \quad |p_j(Aw)| \ge \frac{\eta}{(1 + \|w\|)^{n\bar\delta}} \quad \text{if} \quad j \notin J \right\}$$

It follows that for any z with $\|z\| \gg 0$ there is a unique J such that $A^{-1}(z) \in \mathcal{Z}_J$.

Fix J of cardinal $\#(J) = l$. Then, as a consequence of (5.26), for any z with $w = A^{-1}(z) \in \mathcal{Z}_J$,

$$\max_{j \in J} |\tilde{p}_{J,j}(Aw)| \le \frac{c\eta}{(1 + \|w\|)^{n\bar\delta}} \le \frac{\epsilon}{(1 + \|w\|)^{n\bar\delta}} .$$

Hence, $w \in \mathcal{X}_J$, and so $w \in \mathcal{Y}_l$, which implies that

$$\sum_{j \notin J} |\Lambda_j(w)| \ge \epsilon_0 \|w\| .$$

From this point on, the proof of Proposition 5.7 can be followed verbatim. $\square$

With this proposition at hand, we can propose an analytic method to solve the Bezout identity, with estimates similar to Brownawell's estimate (5.7). Everything in our procedure is explicit, except for the first step, which is based on the pigeonhole principle. We start with m polynomials $P_1, \ldots, P_m$, with degrees $D_1 \ge D_2 \ge \ldots \ge D_m$, without any common zeroes in $\mathbb{C}^n$. We can assume that $m > n$. (If not, repeat the last polynomial.) The first step consists in defining $n + 1$ polynomials q_j, which are linear combinations of the P_l, and $\deg(q_j) = D_j$, $1 \le j \le n$. (The coefficients can be taken to be integers of absolute value bounded by $D_1 \cdots D_n$ [51].) We can now apply the last proposition to the system $q_1, \ldots, q_n$, and find a normal system $p_1, \ldots, p_n$. (Here, the absolute value of the integral coefficients of the linear combinations defining the p_j, are bounded by $(D_1 + 1)^{n-1}$ [12, Lemma 5.2].) As soon as one has some estimate for the constant K, and the choice of coordinates, that appear in the geometric localization (5.18) of all the varieties W_J, one can find linear forms $L_1, \ldots, L_n$ such that (5.25) holds. We now choose a linear combination, p_{n+1}, of all the original polynomials P_j, such that p_{n+1} does not vanish on the set $\{p_1 L_1 = \ldots = p_n L_n = 0\}$. The condition (5.25) implies that if $N > 1$, the polynomials $L_j^{Nn\delta} p_j$, $1 \leq j \leq n$, satisfy the condition (5.8) for

$$d = (N-1)n\delta \quad \text{where} \quad \delta = D_1 \cdots D_n. \tag{5.27}$$

We are left with the task of choosing N large enough to ensure condition (5.9) holds for the smallest integer q possible. In fact, the integer q appearing in (5.10) has to satisfy

$$(q+1)n(N-1)\delta \geq 2(n-1)(D_1 + n\delta) + 1. \tag{5.28}$$

For instance, if we take $N = 2n + 1$, the last inequality holds for $q = 0$. The division formula (5.10), with p_j in the place of P_j, solves now the Bezout equation (5.1), with the estimate

$$\deg(A_j) \leq n(D_1 + nN\delta) \leq 4n^3 \delta. \tag{5.29}$$

With the preceding notation, the Bezout equation becomes

$$1 = < \left(\bar{\partial} \left(\frac{1}{L_1^{nN\delta} p_1} \right) \wedge \cdots \wedge \bar{\partial} \left(\frac{1}{L_n^{nN\delta} p_n} \right) \right)(\zeta), \frac{\Delta_N(z,\zeta)d\zeta}{p_{n+1}(\zeta)} >, \tag{5.30}$$

where Δ_N is the determinant defined as Δ in Proposition 5.4, with the polynomials $L_j^{nN\delta} p_j$, $1 \leq j \leq n$, and p_{n+1} (and their corresponding Hefer polynomials), instead of $P_1, \ldots, P_{n+1}$.

Clearly, the bounds (5.29) we obtain by this scheme are not sharp. In fact, that is also the case for the Brownawell bounds (5.7), due to one of the factors n, which is intrinsic to Skoda's method, and another n, from the Lojasiewicz-Brownawell inequality (5.6). The only way to remove this relatively minor blemishes is to give a geometric proof of the Nullstellensatz. (The counterpart is that the geometric proof will be less constructive.) In fact, J. Kollár [48] proved the following beautiful result, which we state in a version due to Brownawell [25].

Theorem 5.10. *Let $\mathbb{K}$ be an arbitrary field, I an ideal in $\mathbb{K}[X_1, \ldots, X_n]$, generated by m polynomials $P_1, \ldots, P_m$ such that $\deg(P_j) \leq D_j$, with*

$$D_2 \geq \ldots \geq D_m \geq D_1 > 0, \quad D_m \geq 3. \tag{5.31}$$

Let $\mathcal{Q}_1, \ldots, \mathcal{Q}_r$ be the prime minimal ideals in the decomposition of I in primary components; then there are integers $e_1, \ldots, e_r$ such that

$$\mathcal{Q}_1^{e_1} \cdots \mathcal{Q}_r^{e_r} \subseteq I, \quad e_1 + \cdots + e_r \leq D_1 \cdots D_\mu, \quad \mu = \min(n, m). \tag{5.32}$$

When the base field is infinite, Theorem 5.10 implies that, if (5.31) holds, the Bezout equation has solutions A_j such that

$$\deg(A_j) \le D_1 D_2 \cdots D_\mu . \tag{5.33}$$

It is possible to eliminate the condition (5.31) by paying some penalty on the bounds (5.33). Theorem 5.10 has also consequences for the membership problem, as we will mention in the next section, as well as for the Lojasiewicz inequalities [46] (which allows to drop the factor n, provided (5.31) is fulfilled.) There are different versions of Theorem 5.10, due to Heintz, Philippon, and others, and we refer to the excellent survey [71] for the exact references and relationships among them.

In the earlier discussion of Brownawell's estimates, we pointed out that the two parasite factors n appeared because of two different reasons. One seems to be eliminated through the algebraic geometric method of Kollár, the other one seems to be inherent to the analytic tools. In fact, as we have mentioned before, analytic division formulas provide explicit versions of the global Briançon-Skoda theorem. When the algebraic variety associated to the ideal $I(P_1, \ldots, P_m)$ is empty, one cannot differentiate the Briançon-Skoda theorem from solving the Bezout identity. As we know, Briançon-Skoda has an analogue in regular local rings, the Lipman-Teissier theorem. In order to use this analogy, one needs to consider the Bezout equation, or more generally, the membership problem, when the coefficients are not any longer in a field but in a ring. The simplest example, which is also very important in complexity estimates for computer algebra, occurs when the polynomials have integer coefficients. The following result was obtained by Philippon as a consequence of the Lipman-Teissier theorem [63].

Theorem 5.11. *Let $P_1, \ldots, P_m$ have coefficients $\pi_{j,k}$ in $\mathbb{Z}$, whose degrees satisfy the condition (5.31), and such that the naive heights satisfy*

$$h_{P_j} := \max_k \log |\pi_{j,k}| \le h_P := \max_j h_{P_j} \le h . \tag{5.34}$$

Assume further that there are no common zeroes in $\mathbb{C}^n$. Then, one can find $a \in \mathbb{N}^*$ and polynomials $A_1, \ldots, A_m \in \mathbb{Z}[X_1, \ldots, X_n]$ such that

$$a = A_1 P_1 + \cdots + A_m P_m ,$$

$$\log(a) \le \kappa(n) h D_1 \cdots D_\nu \left(\frac{1}{h} + \sum_{j=1}^{\nu} \frac{1}{D_j} \right) , \tag{5.35}$$

$$\deg(P_j A_j) \le (n+2) D_1 \cdots D_\mu , \tag{5.36}$$

where $\mu = \min(n, m)$, $\nu = \min(n+1, m)$, and $\kappa(n)$ is an explicit constant depending only on the dimension n.

The presence of the factor n (really, $n + 2$) in (5.36), is now almost unavoidable, it is a consequence of the use of the Lipman-Teissier theorem. This theorem holds also when $\mathbb{Z}$ is replaced by $\mathbb{Z}[T_1, \ldots, T_k] = \mathbb{Z}[T]$, as it has been shown in [12, Lemma 4.3]. Namely, let $P_1, \ldots, P_m$ be polynomials in $\mathbb{Z}[T][X]$, which have no common zeroes in $\mathcal{Z} \times \mathbb{C}^n$, where $\mathcal{Z}$ is an algebraic closure of $\mathbb{Z}(T_1, \ldots, T_k)$; denote as $D(P)$ (resp. H_P) the maximum of their degrees (resp., their naive heights) when considered as polynomials in all variables (T, X). Let as before $\mu = min(m, n)$; one can find in the ideal generated by the P_j in $\mathbb{Z}[T, X]$ a polynomial $a(T)$, depending only of the variables T, and such that

$$\deg(a) \leq \kappa(n)(D(P))^\nu \quad \text{and} \quad h_a \leq \kappa(n)D^\nu(H_P + (n+k)log(D(P)+1). \quad (5.37)$$

Such a result, when applied to a collection of n polynomials with integral coefficients and degree less than D, defining a discrete variety in $\mathbb{C}^n$, provides a list of polynomials of a single variable $Q_1(X_1), \ldots, Q_n(X_n)$, which lie in the ideal generated by the P_j in $\mathbb{Z}[X_1, \ldots, X_n]$ (namely, $Q_j = \sum_{k=1}^m A_{j,k}P_j$) and have degree and logarithmic size estimates of the form (5.37) (here $\mu = n$.) With such a result, our proposition 4.11 becomes in some sense effective.

Let us give here some insight to this question. If R is any polynomial with integral coefficients, it follows from the transformation law that

$$< \bar{\partial}(1/P), R \, d\zeta > = < \bar{\partial}(1/Q), \Delta R \, d\zeta >, \quad (5.38)$$

where Δ is the determinant of the matrix $[A_{j,k}]$. We know that the global residue in (5.38) is a rational number. One sees that, as a denominator for it, one can choose a common denominator for all numbers

$$< \bar{\partial}(1/Q)(\zeta), \zeta^{*k} d\zeta >, \quad |k| \leq \deg(R) + n \max_{j,k} \deg(A_{j,k}).$$

Such a denominator can be computed using the fact that the Q_j are single variable polynomials for which we know estimates for the coefficients. In fact, it may be more convenient, since one knows that the order of the residue current does not exceed the multiplicity, to compute a denominator δ_α for each local residue (considered as an algebraic number)

$$< \bar{\partial}(1/Q)(\zeta), (\zeta - \alpha)^k \Delta d\zeta >_\alpha,$$

where α is a common zero of the P_j, and $|k|$ does not exceed the multiplicity of α, that is, at most D^n; to do that, it is enough to compute a denominator δ_α for all the residues

$$< \bar{\partial}(1/Q)(\zeta), (\zeta - \alpha)^{k'} d\zeta >_\alpha, \quad |k'| \leq D^n + n\kappa(n)D^{\mu+1}.$$

We can choose as a denominator for (5.38) the product of all δ_α (there are no more than D^n factors because of Bezout's theorem.) The same reasoning works when R is a rational function. Once a denominator has been found, an estimate for the numerator just follows from the fact that we know a priori that the number is rational and that we can perform analytic estimates for (5.38), just using Lojasiewicz-Brownawell inequalities jointly with Bochner-Martinelli integral formulas such (3.32) or (3.33).

Such estimates can be carried on within the context of our representation formula (5.30). In this formula, the degrees of the polynomials appearing in the residue current have degrees approximately equal to δ, this would a priori lead to estimates for the logarithms of the denominators (of the coefficients of A_j) of the form $\kappa(n)h\delta^n$ (h is a bound for the naive height of the p_j, $j \leq n$.) Nevertheless, because the number of *distinct* points contributing to local residues is $\sim \delta$, one can deduce from this formula the following bounds for solutions of the Bezout identity [12, Theorem 5.1].

Theorem 5.12. *Let* $P_1, \ldots, P_m \in \mathbb{Z}[X_1, \ldots, X_n]$, *of degrees at most D and naive heights bounded by h, without common zeroes in $\mathbb{C}^n$. Then, there is a positive integer a and polynomials $A_1, \ldots, A_m$ in $\mathbb{Z}[X_1, \ldots, X_n]$, such that*

$$\sum_{j=1}^{m} A_j P_j = a,$$

with the a priori estimates

$$\deg(A_j) \leq n(2n+1)D^n,$$
$$\max(\log(a), \max_j h_{A_j}) \leq \kappa(n)D^{8n+3}(h + \log m + D \log D).$$

Remark 5.13. The degree estimates are better than those after (5.30), this is due to the fact that we used the Ji-Kollár-Shiffman inequalities, instead of the Lojasiewicz-Brownawell inequalities, this requires the condition $D \geq 3$, which we omitted from the hypotheses for simplicity. Furthermore, one can replace D^n by $D_1 \cdots D_\mu$, with the same notations as in Theorem 5.11, throughout this theorem [38].

To conclude this section, we would like to emphasize here the following idea; it is clear that the Grothendieck residue plays a role not only to control intersection problems (that is, more concretely, it provides bounds for the denominators in division formulas,) but also, simultaneously, it provides estimates for the size of the quotients. This was already clear in the formula (4.54), which we gave to prove Macaulay's theorem.

§2 The membership problem

As we mentioned at the beginning of the previous section, an effective solution to the algebraic Nullstellensatz can be obtained from a solution of the Bezout identity. This is the well-known Rabinowitsch trick [76], which consists in introducing an extra variable. It follows from Brownawell's or Kollár's results, Theorem 5.10, that if an ideal I is generated by polynomials $P_1, \ldots, P_m \in \mathbb{C}[X]$, satisfying the hypotheses (5.31), and if Q vanishes on the variety $V \subset \mathbb{C}^n$ of common zeroes of the P_j, then one can write

$$Q^{D_1 \cdots D_\mu} = \sum_{j=1}^{m} B_j P_j, \quad \deg(B_j) \leq (1 + \deg(Q)) D_1 \cdots D_\mu, \ \mu = \min(m, n).$$

(5.39)

More difficult is the problem of writing explicitly Q in the ideal, when one already knows that $Q \in I$. Note that an intermediate situation between an effective solution for the membership problem and an effective one for the Nullstellensatz, is to solve, with good bounds, the equation

$$Q^s = \sum_{j=1}^{m} B_j P_j,$$

(5.40)

for some power s, s depending only on n. For instance, from Briançon-Skoda, one knows that if Q belongs to the integral closure $\bar{I}$, then (5.40) has a solution for $s = \min(n, m)$. For effective testing of this kind of problems in computer algebra, it is very useful to have a priori bounds on the degrees, heights, etc., of the B_j. In the membership problem, one would like to solve (5.40) with $s = 1$ and at the same time have a priori estimates. Unfortunately, this is not possible in general. There is an example of Mayr-Meyer [52],[6], where for each integer $D \geq 5$, each integer $k > 1$, and letting $n = 10k$, one can find polynomials $F_1, \ldots, F_{n+1}$, of n variables, with integral coefficients (in fact, differences of monomials), such that X_1 belongs to the ideal $I(F_1, \ldots, F_{n+1})$, but any solution A_j for the membership equation

$$X_1 = \sum_{j=1}^{n+1} A_j F_j,$$

satisfies

$$\max_j \deg(A_j) \geq (D - 2)^{2^{k-1}}.$$

Because the Buchberger algorithm is so powerful that it solves simultaneously many problems associated to a given ideal, one can use this example to show that its complexity is doubly exponential [6].

One can obtain better bounds under special hypotheses. For instance, if the $P_1, \ldots, P_m$ define a complete intersection, then one can easily prove using L^2 estimates for the $\bar{\partial}$-operator [11], that if $Q \in I$, one can solve

$$Q = \sum_{j=1}^{m} A_j P_j,$$

(5.41)

with bounds

$$\deg(A_j) \le \deg(Q) + \kappa(n)D_1 \cdots D_m \,. \tag{5.42}$$

Since all prime ideals in the decomposition of an ideal defined by a regular sequence are isolated, this result also follows from Theorem 5.10, which additionally improves on (5.42), namely, $\kappa(n) = 1$, provided the condition (5.31) is fulfilled. Let us now show how the Grothendieck residue can be used to solve this problem, furnishing also bounds for the heights. We shall use the notation from (4.23): for a family of functions $f_1, \ldots, f_r$ and $S \subseteq \{1, \ldots, r\}$, we denote $F_S = \prod_{j \in S} f_j$. (Note that $F_{\{j\}} = f_j$.)

Proposition 5.14. *Let $f_1, \ldots, f_n$ be n polynomials of degree at most D, satisfying the properness condition (4.51), with constants $\gamma, d > 0$. Let $m \le n$ and $h \in I(f_1, \ldots, f_m)$, then, for any integer q such that*

$$qd > \deg(h) + (D - d)(3n - 1) + n - D \,,$$

we have the identity

$$h(z) = \sum_{s=1}^{m} \sum_{\substack{S \subseteq \{1,\ldots,m\} \\ \#(S)=s}} \left(\sum_{|k| \le q} c_{S,k} f^{*k}(z) \right) F_S(z) \,, \tag{5.42}$$

where, letting $\vartheta = (2, \ldots, 2, 1, \ldots, 1)$, a multiindex of n components, whose first m components are twos,

$$c_{S,k} = \left\langle \bar{\partial}\left(\frac{1}{f^{*(k+\vartheta)}} \right)(\zeta), h(\zeta) \left(\prod_{\substack{1 \le j \le m \\ j \notin S}} (f_j(\zeta) - f_j(z)) \right) g(z, \zeta) \right\rangle \,. \tag{5.43}$$

Proof. We choose $\epsilon_j \sim N$ so that the set $\Delta_f(\epsilon) = \{|f_j| < \epsilon_j : 1 \le j \le n\}$ is a Weil polyhedron containing all the common zeroes of $f_1, \ldots, f_n$. Inside this polyhedron we apply the formula (4.27) to represent the polynomial h (use $p = m$ in that formula.) We recall that the last integral in (4.27) is zero because h is in the ideal. Developing all the integrands that appear in the formula as geometric series, as we usually do in the Cauchy-Weil formula, we obtain in $\Delta_f(\epsilon)$ that

$$h(z) = \sum_{s=1}^{m} \sum_{\substack{S \subseteq \{1,\ldots,m\} \\ \#(S)=s}} \left(\sum_{k \in \mathbb{N}^n} c_{S,k} f^{*k}(z) \right) F_S(z) \,. \tag{5.44}$$

We apply Theorem 4.13 to conclude that $c_{S,k} = 0$ for

$$(|k| + 2n)d > \deg(h) + D(2n - s) + (n - 1)(D - d) + n \,,$$

so that the series (5.44) has only non trivial terms for $|k| \le q$. $\qquad\square$

Let us indicate how this proposition can be used to solve effectively the membership problem in the case of complete intersection [39]. Assume that $P_1, \ldots, P_m$ define a complete intersection in $\mathbb{C}^n$, and denote by δ the product of their degrees. Adapting the proof of Proposition 5.8 and using the Lojasiewicz-Brownawell inequality (5.6), one can construct m linear combinations p_j of the P_j, with an invertible matrix with integral coefficients, n affine forms $L_1, \ldots, L_n$, and $n - m$ linear forms $p_{m+1}, \ldots, p_n$ such that, for any $N \in \mathbb{N}^*$, there exists a positive constant γ_N with the property that, if $\|z\| \gg 0$,

$$\max_{1 \le j \le n} |L_j(z)^{nN\delta} p_j(z)| \ge \gamma_N \|z\|^{n(N-1)\delta}. \tag{5.45}$$

Moreover, the choice of the L_j is generic, that is (5.45) holds for a collection of $n + 1$ families $\{L_j^{(\tau)}\}$ (indexed by τ) of n linear forms each, which can be chosen such that the different polynomials

$$\Phi^{(\tau)} = \prod_{1 \le j \le m} L_j^{(\tau)}$$

have no common zeroes in $\mathbb{C}^n$. We use the division formula (5.42) with $f_j = L_j^{Nn\delta} p_j$ to represent, in the ideal generated by $f_1, \ldots, f_m$, the polynomial $h = \Phi^{Nn\delta} Q$, where we have fixed τ and set $\Phi = \Phi^{(\tau)}$. The multiplication by Φ is necessary to guarantee that h belongs to that ideal. Choosing, for example, $N = 2$, the value of q terminating the series is roughly $8n + (\deg(Q)/n\delta)$, so that

$$\Phi^{2n\delta} Q = \sum_{j=1}^{m} A_j f_j = \sum_{j=1}^{m} B_j p_j .$$

The polynomials $B_j = B_j^{(\tau)}$ have the degree estimate

$$\deg(B_j) \le \text{const.} \ n^2 \delta + 2 \deg(Q)$$

Due to the choice of $n + 1$ families of linear forms, we can solve the Bezout equation for the $\Phi^{(\tau)}$ and then raise both sides of the identity to the power $2(n+1)n\delta$, so that one can find polynomials C_j such that

$$Q = \sum_{j=1}^{m} C_j p_j , \quad \deg(C_j) \le 2 \deg(Q) + \kappa(m, n)\delta .$$

Here $\kappa(m, n) \le \text{const.} \ n^4 m^n$. These estimates are not sharp, but their interest lies in the fact that, again, we can use this construction for polynomials with integral coefficients and obtain estimates for the naive height of the C_j similar to those in Theorem 5.12.

To continue the discussion of the membership problem in the case of a complete intersection, and to support our remarks of the end of §1, let us explain the relationship between multidimensional residues and the Chow form of an ideal. We recall from [44],[54],[55], that the Chow form of an unmixed homogeneous ideal $\mathcal{I}$ of rank $n + 1 - d$ in $\mathbb{F}[X_0, \ldots, X_n]$ ($\mathbb{F}$ subfield of $\mathbb{C}$, as before), is defined as follows. Consider linear forms $L_1, \ldots, L_d$ of $n+1$ variables, depending linearly on a set of parameters U, namely,

$$L_j(X) = L_j(X; U) = U_{j,0} X_0 + \cdots + U_{j,n} X_n.$$

Let $\mathcal{G}$ be collection of all polynomials $G \in \mathbb{F}[U]$ with the property that for some integer $M \geq 1$, we have the inclusion of ideals

$$(X_0, \ldots, X_n)^M G \subseteq (\mathcal{I}, L_1, \ldots, L_d).$$

$\mathcal{G}$ is a non-zero principal ideal in $\mathbb{F}[U]$ and any generator of $\mathcal{G}$ is called a Chow form (or an eliminating form in the terminology of [60].) We can now state the following proposition.

Proposition 5.15. *Let* $P_1, \ldots, P_m$ *be* m *homogeneous polynomials of* $n + 1$ *variables, with coefficients in* $\mathbb{F}$ *and respective degrees* D_j, *defining a complete intersection in* $\mathbb{C}^{n+1}$. *Let* $L_{m+1}, \ldots, L_{n+1}$ *be* $n + 1 - m$ *linear forms, of the form (5.46) and let* $\tilde{P} = (P_1, \ldots, P_m, L_{m+1}, \ldots, L_{n+1})$. *Then, there is a family of rational functions* $\mathcal{R}_{(k,l)} \in \mathbb{F}(U)$, *indexed by* $(k, l) \in (\mathbb{N}^{n+1})^2$, *subject to the conditions*

$$\sum_{j=1}^{m} k_j D_j + \sum_{j=m+1}^{n+1} k_j \leq D_1 + \cdots + D_m + 1 - m, \tag{5.46}$$

$$|l| \leq \sum_{j=1}^{m} (k_j + 1) D_j + \sum_{j=m+1}^{n+1} k_j - m, \tag{5.47}$$

such that for any test function ϕ *in* $\mathbb{C}^{n+1}$,

$$< \bar{\partial}\left(\frac{1}{\tilde{P}^{k+\underline{1}}}\right), \phi \, d\zeta > = \sum_{l} \mathcal{R}_{k,l} \frac{\partial^{|l|}}{\partial \zeta^l} \phi(0). \tag{5.48}$$

Any common denominator for the $\mathcal{R}_{k,l}$ *is a multiple of the Chow form of the ideal* $\mathcal{I}(P_1, \ldots, P_m)$.

Proof. Since the P_j define a complete intersection, it follows from Proposition 4.8 and Theorem 4.14 that, for any generic choice of the set of parameters U, any test function ϕ, and any index k satisfying (5.46),

$$< \bar{\partial}\left(\frac{1}{\tilde{P}^{k+\underline{1}}}\right), \phi \, d\zeta > = \sum_{l} c_{(k,l)} \frac{\partial^{|l|}}{\partial \zeta^l} \phi(0).$$

The coefficients $c_{(k,l)} = c_{(k,l)}(U)$ are zero if the lenght of l does not satisfy the restrictions (5.47). On the other hand, as an immediate application of the transformation law (see Proposition 4.11, with the field $\mathbb{F}(U)$ instead of $\mathbb{F}$), one can see that the coefficients $c_{(k,l)}$ are in fact rational functions of U. These are the rational functions $\mathcal{R}_{(k,l)}$. The right-hand side of formula (4.56) belongs to $\mathbb{F}(U)$, it admits as a denominator any common denominator for the $\mathcal{R}_{(k,l)}$. Since (4.56) is also valid for all monomials (with κ the degree of the monomial), the definition of a Chow form implies that the last statement of the proposition holds. In fact, careful consideration of the transformation law shows that the $\mathcal{R}_{(k,l)}$ are powers of the Chow form. $\square$

Remark 5.16. The numerators of the same rational functions $\mathcal{R}_{(k,l)}$, play a role in the estimation of the quotients in the membership equation. Moreover, as it is known that when the polynomials P_j have integral coefficients, the height of the Chow form measures the arithmetic complexity of the cycle defined by the P_j [64], it seems that in the case of a complete intersection, the multidimensional residue should play a similar role.

Remark 5.17. The role of the Grothendieck residue in studying the complexity of computer algebra algorithms has also been pointed out by Dickenstein and collaborators in a series of recent papers [31],[32],[41].

We would like now to consider an altogether different approach to membership problems. These are ideas have been mainly introduced to study similar division problems in algebras of entire functions, of which, we shall see an example in the following section. We return to our point of view in Chapter 3, §§4,5, which consisted in writing division formulas containing the distribution-valued function $|f_1|^{2\lambda_1} \cdots |f_p|^{2\lambda_p}$ and its analytic continuation.

In order to study global division problems, instead of semilocal ones, as we did in Chapter 3, we need a weighted Bochner-Martinelli formula for the representation of smooth functions, not necessarily holomorphic. In the following proposition, we return to the notation of Theorem 2.7, except that the kernels will be denoted $\tilde{P}$ and $\tilde{K}$, we assume an additional condition on s. Namely, given the domain U and $\omega \subset\subset U$, we require the existence of a strictly positive constant $\gamma = \gamma(\omega)$ such that

$$| < s, z - \zeta > | \geq \gamma \|z - \zeta\|^2 \quad \forall z \subset \omega, \zeta \in \bar{U}. \tag{5.49}$$

Proposition 5.18. (Koppelman's formula.) *Let U be a bounded domain with piecewise smooth boundary, ϕ a function in $C^1(\bar{U})$, and the notation as in Theorem 2.7. For any $z \in \omega$, one has the representation formula*

$$\phi(z) = \frac{1}{(2\pi i)^n}\left(\int_U \phi(\zeta)\tilde{P}(z,\zeta) + \int_{\partial U} \phi(\zeta)\tilde{K}(z,\zeta) - \int_U \bar{\partial}\phi(\zeta) \wedge \tilde{K}(z,\zeta) \right). \tag{5.50}$$

Sketch of the proof. A complete proof can be found in [17],[18]. A careful look at the kernels $\tilde{K}$ and $\tilde{P}$ shows that, outside the diagonal in $\mathbb{C}^n \times \mathbb{C}^n$, one has

$\tilde{P}(z,\zeta) = -d_\zeta \tilde{K}(z,\zeta)$. Therefore, formula (5.50) can be obtained exactly as we have obtained the standard Bochner-Martinelli formula (2.16). The key remark one should make here, is that the kernel $\tilde{K}$ can be expressed as $\tilde{K} = \tilde{K}_0 + \tilde{K}_1$, where

$$\tilde{K}_0(z,\zeta) = (\Gamma_1^0 \ldots \Gamma_M^0)(z,\zeta) \frac{S \wedge \bar{\partial} S^{n-1}}{<s, \zeta - z >^n}$$

and $\tilde{K}_1$ is sum of the forms with singularities of the form $\|\zeta - z\|^{-2(k+1)}$, $k = 0, \ldots, n-2$, on the diagonal. One can apply Stokes's formula to the domain $U \setminus \bar{B}(z, \epsilon)$ and let $\epsilon \to 0$ to obtain (5.50). $\square$

The other ingredient that we need to introduce are the Bernstein-Sato functional equations. Consider a family of polynomials $P_1, \ldots, P_M$ in n variables, with coefficients in a field $\mathbb{F}$ of characteristic zero. Let $\lambda_1, \ldots, \lambda_M$ be a set of parameters. We claim that there exists a collection of differential operators $\mathcal{Q}_1, \ldots, \mathcal{Q}_M$ with coefficients in $\mathbb{F}[\lambda, X]$ (acting on the variables X) and a polynomial B in $\mathbb{F}[\lambda]$ such that one has the following set of formal identities,

$$\mathcal{Q}_j(P_j P^{*\lambda}) = B(\lambda) P^{*\lambda}, \quad 1 \le j \le M. \tag{5.51}$$

The meaning of these identities is that, after we follow the rules of differentiation of powers, we obtain true polynomial identities in $\mathbb{F}[\lambda, X]$. The existence of such identities was found first by J. Bernstein for $M = 1$ [20], and generalized by Lichtin [49]. The basic idea is to use the fact that any finitely generated, non-trivial, $\mathbb{F}(\lambda) < X, \partial_X >$-module (where $\mathbb{F}(\lambda) < X, \partial_X >$ denotes the Weyl Algebra over the field $\mathbb{F}(\lambda)$) has a homological dimension bigger or equal than n. (see [21, Chapter 1],[22, p. 173].) Holonomic modules are those with homological dimension exactly equal to n. Such is the case for the module

$$\mathcal{N} = \mathcal{M} \oplus \cdots \oplus \mathcal{M} \quad (M \text{ times})$$

where

$$\mathcal{M} = \left\{ \frac{R(\lambda, X)}{P^{*\underline{v}}} P^{*\lambda} : v \in \mathbb{N}, R \in \mathbb{F}(\lambda)[X] \right\}$$

Using a good filtration (indexed on v), one can compute the multiplicity l of the module $\mathcal{N}$ as the coefficient of v^n in

$$Mn! \binom{n + v(\deg(P_1 \cdots P_M) + 1)}{n},$$

which is, roughly, $\kappa(n)(\max \deg(P_j))^n$. Any strictly decreasing sequence of submodules has length at most l. In particular, one can apply this fact to the sequence of modules

$$\mathbb{F}(\lambda) < X, \partial_X > \cdot \left(\frac{P^{*(\lambda+\underline{q})}}{P_1}, \ldots, \frac{P^{*(\lambda+\underline{q})}}{P_M} \right) \quad q \in \mathbb{N}$$

The stationarity of this sequence implies the existence of the system of functional equations (5.51).

There is an analogous result for germs of holomorphic functions, with the algebra of differential operators with holomorphic coefficients replacing the Weyl algebra. Moreover, the polynomial B can be taken to be a product of affine functions with positive integral coefficients [67]. When $M = 1$, $P_1 \in {}_n\mathcal{O}_0$ fixed, the set of all the polynomials B appearing in an equation of the type (5.51) constitutes a principal ideal, with a generator that has strictly negative rational roots [47]. (It is an interesting question whether this property holds for polynomials P_1 over a field $\mathbb{F}$ of characteristic zero.)

For a differential operator R with polynomial coefficients in $\mathbb{F}[X]$

$$R(X, \partial_X) = \sum_\alpha \rho_\alpha(X) \frac{\partial^{|\alpha|}}{\partial X^\alpha}$$

we define the defect $e(R)$ as

$$e(R) = \max_\alpha(\deg(\rho_\alpha) - |\alpha|)$$

For a fixed polynomial $P \in \mathbb{F}[X_1, \ldots, X_n]$, we introduce the curious notion of defect, $E(P)$, as follows: we express any functional equation (5.51) valid for P as

$$\sum_l \lambda^l R_l(X, \partial_X) P^{\lambda+1} = B(\lambda) P^\lambda$$

Then, the defect $\tilde{e}(R(\lambda, X, \partial_X)) := \max_l e(R_l)$ and, finally, $E(P)$ is the minimum of $\tilde{e}(R(\lambda, X, \partial_X))$ for all possible functional equations (5.51) for P. It is not hard to verify that

$$E(P) \geq -\deg(P)$$

The following result relates these concepts to the membership problem [7, Theorem 4.1].

Theorem 5.19. *Let $P_1, \ldots, P_m$ be polynomials defining a complete intersection variety V in $\mathbb{C}^n$, of degree not exceeding D. Let*

$$E = \max\{E(P^{*k}) : k \in \mathbb{N}^n, |k| \leq m\}$$

Then, any polynomial $Q \in I(P_1, \ldots, P_m)$, can be written as

$$Q = \sum_{j=1}^m A_j P_j$$

with

$$\deg(A_j) \leq \deg(Q) + (5m - 2)D + 4E + 4n - 3. \tag{5.52}$$

Remark 5.20. There is a similar result when one drops the hypotheses of complete intersection, for the effective membership of $Q^{\min(n,m)}$ in I, when Q is known to be in the integral closure $\bar{I}$. In this case, the defect that appears in the estimate analogous to (5.52) [15, Proposition 5], is $E(\mathcal{P})$, where $\mathcal{P}$ is the polynomial with real coefficients in $2n$ variables given by

$$\mathcal{P}(X,Y) = \sum_{j=1}^{m} |P_j(X+iY)|^2.$$

As in [16], one can define a multivariable version of defect, $E(P_1, \ldots, P_M)$, taking into account all possible systems of functional equations (5.51). Then, using $E(P_1, \ldots, P_m)$, one can give estimates of the type (5.52), so that the underlying complex structure of the problem is not lost.

Proof of Theorem 5.19. The proof of this theorem, as well as that of the last remark, is quite technical, but it may be adaptable to many division problems in spaces of holomorphic functions, like in [16]. We shall give a brief description of the proof and we refer the reader to [7, §4], [16, §3], and [15, Proposition 5], for the details. We can assume that $P_1, \ldots, P_m$ define a normal system (see proof of Proposition 5.8.) We use Proposition 5.18 with two weights. For the notation, we refer to the proof of Theorem 2.7. In what follows, N will be a sufficiently large integer that will be chosen at the end of the proof. The first weight we consider is constructed starting with a polynomial $\Gamma_1(t)$ of a single variable, $\Gamma_1(1) = 1$, and a smooth $(1,0)$ differential form Q_1 in $\mathbb{C}^{2n}$, depending on a single complex parameter μ (or, if necessary, on several complex parameters λ.) For example, under the hypotheses of Theorem 5.19, one uses one complex parameter and chooses, as in the proof of the Local Duality Theorem,

$$Q_1(z,\zeta;\mu) := \frac{1}{m} \sum_{j=1}^{m} |P_j(\zeta)|^{2\mu} \frac{g_j(z,\zeta)}{P_j(\zeta)}$$

$$\Gamma_1(t) := \frac{1}{m!} \prod_{j=0}^{m-1} (mt - j)$$

The second pair is given by

$$Q_2(z,\zeta) := \partial \log(1 + |\zeta|)^2$$

$$\Gamma_2(t) := t^N$$

To every pair we associate a function as before,

$$\Phi_j(z,\zeta) := 1 + <Q_j, z - \zeta> := 1 + \sum_{k=1}^{n} q_{j,k}(z,\zeta)(z_k - \zeta_k),$$

A simple computation shows that

$$\Phi_1(z,\zeta;\mu) = \frac{1}{m}\sum_{j=1}^{m}|P_j(\zeta)|^{2(\mu-1)}\overline{P_j(\zeta)}P_j(z) + \frac{1}{m}\sum_{j=1}^{m}(1-|P_j(\zeta)|^{2\mu})$$

$$\Phi_2(z,\zeta) = \frac{1+<z,\bar{\zeta}>}{1+\|\zeta\|^2}.$$

We choose a radial function $\chi \in C_0^\infty(\mathbb{C}^n)$, $\chi \equiv 1$ for $\|\zeta\| \leq 1$, $\chi \equiv 0$ for $\|\zeta\| \geq 2$, $0 \leq \chi \leq 1$. For a fixed $R > 1$, apply the representation formula (5.50) in the ball $B(0, 2R)$ to the function $\phi(\zeta) := \chi(\zeta/R)Q(\zeta)$. Note that (5.50) is a priori defined only when the parameter μ satisfies Re $\mu \gg 1$, and we apply it to a fixed z, $\|z\| < R/2$. The two integrals that remain in (5.50) admit an analytic continuation to the whole complex plane as meromorphic functions of μ. We are going to identify the zeroth coefficient of their Laurent development at $\mu = 0$, which will provide a representation for $Q(z)$.

Following the computations in the proof of Theorem 3.21, we can conclude that because $Q \in I$, the first integral in (5.50) represents an element of the ideal generated by I in $C_0^\infty(\mathbb{C}^n)$. (It is here that one uses the fact that $P_1,\ldots,P_m$ is a normal sequence.) More precisely, if we write

$$\varpi = \frac{1}{m}\sum_{j=1}^{m}(1-|P_j(\zeta)|^{2\mu})$$

we obtain

$$\Gamma_1^{(\alpha_1)}(z,\zeta;\mu) - \frac{d^{\alpha_1}\Gamma_1}{dt^{\alpha_1}}(\varpi) = \sum_{j=1}^{m}P_j(z)\gamma_{\alpha_1,j}(z,\zeta;\mu).$$

We have seen in the proof of Theorem 3.21, that to compute the zeroth coefficient of the first integral in (5.50) at $\mu = 0$, we can replace everywhere in the kernel $\tilde{P}$ the quantity $\Gamma_1^{(\alpha_1)}$ by $\sum_{j=1}^m P_j(z)\gamma_{\alpha_1,j}(z,\zeta;\mu)$.

The other important terms where μ appears are terms of the form $(\bar{\partial}Q_1)^{\alpha_1}$, $\alpha_1 \in \mathbb{N}$, which are linear combination of expressions of the type

$$\mu^{\alpha_1}|P_{j_1}(\zeta)\cdots P_{j_{\alpha_1}}(\zeta)|^{2(\mu-1)}\bigwedge_{l=1}^{\alpha_1}\overline{\partial P_{j_l}(\zeta)} \wedge g_{j_l}(z,\zeta).$$

A typical term in $\Gamma_1^{(\alpha_1)}$ is

$$\left\{\sum_{j=1}^{m}|P_j(\zeta)|^{2(\mu-1)}\overline{P_j(\zeta)}P_j(z) + \sum_{j=1}^{m}(1-|P_j(\zeta)|^{2\mu})\right\}^q, \quad 0 \leq q \leq m-\alpha_1.$$

Finally, both integrals in (5.50) can be replaced by linear combinations of expressions of the following type

$$\frac{\mu^{\alpha_1}}{R^\varepsilon} \int_{\mathbb{C}^n} |P(\zeta)|^{*2(\mu-1)k} |P(\zeta)|^{*2l} Q(\zeta)\chi^{(\varepsilon)}(\zeta/R)\Omega(z,\zeta), \qquad (5.53)$$

where $k, l \in \mathbb{N}^m$, $|k| \le m$, $\chi^{(\varepsilon)}$, the εth derivative of χ, $\varepsilon = 0, 1$, and Ω is a form of degree (n, n), smooth on the support of $\chi^{(\varepsilon)}(\zeta/R)$. The form Ω involves the coefficients of the second pair.

Each expression of the form (5.53), when analytically continued to $\mu = 0$, contributes one term to the zeroth term of the Laurent expansion of (5.50). Let $a_{k;-\alpha_1}$ be the distribution which is the coefficient of $\mu^{-\alpha_1}$ in the Laurent expansion of $|P|^{*2\mu k}$ at $\mu = -1$. Therefore, the contribution of (5.53) is given by expressions of the form

$$< a_{k;-\alpha_1}, \frac{1}{R^\varepsilon}|P|^{*2l}\chi^{(\varepsilon)}(\zeta/R)\omega(z,\zeta) >, \qquad (5.54)$$

where $\omega(z,\zeta)$ is one of the coefficients of $\Omega(z,\zeta)$. Using integration by parts, one can see that the action of the distributions $a_{k;-\alpha_1}$ can be computed with the help of any functional equation (5.51) valid for P^{*k}. We choose the equation that achieves the defect E. We can adjust the choice of N so that the terms corresponding to $\varepsilon = 0$ have a limit when $R \to \infty$ and those corresponding to $\varepsilon = 1$ vanish when $R \to \infty$. The choice of N depends on E and the degrees of expressions involving the polynomials P_j and their Hefer polynomials in (5.54). Letting $R \to \infty$, we obtain the global division formula with the claimed estimates. □

For the Briançon-Skoda problem, the proof is the same, except that one makes another choice of the pair q_1, G_1, namely, $G_1(t) = t^{\min(m,n+1)}$ and there are two natural possibilities for q_1; one is to take one complex parameter μ and define

$$\tilde{q}_1(z,\zeta;\mu) = \|P\|^{2(\mu-1)}\Big(\sum_{j=1}^m \overline{P_j(\zeta)}\vec{g}_j(z,\zeta)\Big)$$

or take m complex parameters λ and let

$$\tilde{q}_1(z,\zeta;\lambda) = |P|^{*2\lambda}\frac{\sum_{j=1}^m \overline{P_j(\zeta)}\vec{g}_j(z,\zeta)}{\|P\|^2}.$$

In the first case, we have to deal with the analytic continuation of the function of two n real variables $\mathcal{P}$, in the second case, one has to get some control on the analytic continuation, as a distribution valued function, of

$$\lambda \mapsto \frac{|P|^{*2\lambda}}{\|P\|^{2n}}.$$

A clever argument based on properties of the Euler Beta function, reduces this problem to estimates involving the multidefect.

To conclude this section, let us return to the question, raised after (5.40), about the best power s one could expect to use in order to decompose Q^s in the ideal $I(P_1, \ldots, P_m)$, with good estimates for the degrees of the quotients, i.e., estimates of the type $D_1 \cdots D_\mu$, $\mu = \min(m, n)$, when we know a priori that Q is in I or in $\bar{I}$. We have shown elsewhere [11] that when the variety V is discrete, then $s = 1$ works for $Q \in I$. When V is a curve, Amoroso has shown that $s = n$ works when $Q \in \bar{I}$ [2]. There are heuristic reasons, arising from the ideas explained in this section, which indicate that $s = n^2$ should always work for $Q \in \bar{I}$. One of them is the close relation between the existence of good bounds and some reasonable control on the order of the distribution $\|P\|^{-2n}$ (considered as a principal value distribution obtained via analytic continuation.) Another one is the power of Lipman-Teissier's theorem, which was already revealed in §1. Thanks to this theorem, jointly with Kollár's method, Amoroso has announced very recently in [3] that $s = 3^n + n^2$ works for any $Q \in \bar{I}$ (the 3^n seems to be an unavoidable consequence of Kollár's hypothesis (5.31).)

§3 The Fundamental Principle of L. Ehrenpreis

L. Ehrenpreis [37] and V. Palamodov [56] have shown that, when Ω is a convex open subset of $\mathbb{R}^n$, one can find a representation of every solution, in $C^\infty(\Omega)$, of a homogeneous system of partial differential equations with constant coefficients, in terms of the exponential polynomial solutions of the same system. The existence of this representation, which extends to $n > 1$ the well known result of Euler for ordinary differential equations, is usually called the Fundamental Principle. As an immediate consequence, the set

$$\mathcal{K} := \{ f \in C^\infty(\Omega), \ P_1(i\frac{\partial}{\partial x}) f = \ldots = P_m(i\frac{\partial}{\partial x}) f = 0 \}, \qquad (5.55)$$

is spanned by the exponential polynomials it contains. Here, the polynomials P_j are the symbols of the partial differential operators. When $\Omega = \mathbb{R}^n$, from the point of view of harmonic analysis, the density of the exponential polynomial solutions corresponds to the fact that the translation invariant space $\mathcal{K}$ admits spectral synthesis [45]. The space of solutions of a finite number of convolution equations (convolution with compactly suported distributions) fails, in general, the property of spectral synthesis [42], though this is true for slowly decreasing systems of convolution equations [9], (see also the survey [8].)

We give in this section a version of the Fundamental Principle, which provides an *explicit* representation of all smooth solutions in a convex open subset of $\mathbb{R}^n$ of a system of partial differential equations with constant coefficients of the form

$$P_1(i\frac{\partial}{\partial x})f = \ldots = P_m(i\frac{\partial}{\partial x})f = 0. \tag{5.56}$$

We are going to make several simplifying assumptions. The first is that the polynomials $P_1, \ldots, P_m \in \mathbb{C}[X_1, \ldots, X_n]$ are in normal position. The second is the following. Let $U \subseteq \mathbb{R}^n$ be a domain, and assume throughout this section that f is an element of $C^\infty(U)$, which satisfies the equations (5.56) in U. Our goal is to represent f in some strictly convex open set $\Omega \subset\subset U$ with smooth boundary. Choose a function $\chi \in \mathcal{D}(U)$, identically equal to 1 in a neighborhood of $\bar{\Omega}$. Using Plancherel's theorem, for any $\phi \in \mathcal{D}(\Omega)$, we have

$$\int_{\mathbb{R}^n} f(x)\phi(x)dx = \int_{\mathbb{R}^n} f(x)\chi(x)\phi(x)dx = \frac{1}{(2\pi)^n}\int_{\mathbb{R}^n} \widehat{f\chi}(\xi)\hat{\phi}(\xi)d\xi. \tag{5.57}$$

By the Paley-Wiener theorem, the entire function $\hat{\phi}$ satisfies in $\mathbb{C}^n$ the estimates

$$\forall m \in \mathbb{N}, \exists c_m > 0 : |\hat{\phi}(\zeta)| \leq \frac{c_m}{(1+\|\zeta\|)^m}e^{H_\Omega(\mathrm{Im}\ \zeta)}, \tag{5.58}$$

where

$$H_\Omega(\xi) = \max_{x \in \bar{\Omega}} < x, \xi >, \ \xi \in \mathbb{R}^n. \tag{5.59}$$

It is clear that

$$\Omega = \{x \in \mathbb{R}^n : \forall \xi, \|\xi\| = 1, \sum_{j=1}^n x_j\xi_j < H_\Omega(\xi)\}.$$

Note that if we define

$$\varphi : \mathbb{C}^n \longrightarrow \mathbb{R}, \quad \varphi(\zeta) = H_\Omega(\mathrm{Im}\ \zeta),$$

the map

$$\zeta \mapsto \mathrm{grad}(\varphi)(\zeta)$$

maps $\mathbb{C}^n \setminus \{\mathrm{Im}\ \zeta = 0\}$ onto $\partial\Omega$.

We represent $\hat{\phi}$ in $\mathbb{C}^n$ with the help of the Koppelman formula (5.50), with a convenient choice of weights. In fact, for any $\psi \in \mathcal{D}(\mathbb{C}^n)$ such that $\psi \equiv 1$ in a neighborhood of a given point z, we have

$$\hat{\phi}(z) = \frac{1}{(2\pi i)^n}\int_{\mathbb{C}^n} \hat{\phi}(\zeta)\psi(\zeta)\tilde{P}(z, \zeta) + \frac{1}{(2\pi i)^n}\int_{\mathbb{C}^n} \hat{\phi}(\zeta)\bar{\partial}\psi(\zeta) \wedge \tilde{K}(z, \zeta). \tag{5.60}$$

We construct $\tilde{K}$ and $\tilde{P}$ as in Theorem 2.7 with $M = 3$. We will express the weights q_l in terms of their associated differential forms. First, we choose a pair, depending on a complex parameter μ, adapted to the division problem. Let

$$Q_1(z, \zeta; \mu) = \frac{1}{m} \sum_{j=1}^{m} |P_j(\zeta)|^{2(\mu-1)} \overline{P_j(\zeta)} g_j(z, \zeta),$$

where $g_{j,k}$ are the Hefer polynomials associated to the P_j. As usual, we take

$$G_1(t) = \frac{1}{m!} \prod_{j=0}^{m-1} (mt - j).$$

We also introduce Q_2 as

$$Q_2(z, \zeta) = Q_2(\zeta) = \partial \log(1 + \|\zeta\|^2),$$

and $G_2(t) = t^N$, N is a positive integer to be chosen later. Finally, we define

$$Q_3(z, \zeta) = Q_3(\zeta) = 2\partial(\varphi * \sigma)(\zeta),$$

where σ is a function in $\mathcal{D}(\mathbb{C}^n)$, supported by the ball $B(0,1)$, nonnegative and with total mass 1. (The reason for convolving with σ is the lack of regularity for the function φ.) To simplify the notation, we shall write $\tilde{\varphi} := \varphi * \sigma$. Associated with Q_3, we let $G_3(t) = e^{t-1}$.

We follow here the proof we already gave for the solution of the algebraic membership problem (Theorem 5.19.) We apply formula (5.50) to $\psi(z) = \theta(z/R)$, where $R > 0$ and θ is some function in $\mathcal{D}(\mathbb{C}^n)$, with support in $B(0,2)$, identically equal to 1 in some neighborhood of $B(0,1)$. Formula (5.50) is valid whenever $\|z\| < R$. The problem, which remains here, is to get rid of the complex parameter μ. This can be done as in the proof of Theorem 5.19, if one remembers that, for any polynomial P, for any test function ω in $\mathcal{D}(\mathbb{C}^n)$, the analytic continuation of

$$\mu \mapsto \int |P(\zeta)|^{2\mu} d\bar{\zeta} \wedge d\zeta, \tag{5.61}$$

exists as a meromorphic function. In fact, the existence of the analytic continuation of (5.61), already obtained as a consequence of Atiyah's theorem, can also be deduced from the algebraic result of J.Bernstein [20] mentioned in the preceding section, namely the existence of a formal functional identity (5.51). This identity becomes a true identity between distributions, if one takes advantage of the fact that ∂ and $\bar{\partial}$ commute and iterates (5.51), thus obtaining the composition of operators

$$\mathcal{Q}(\mu, \zeta, \partial_\zeta) \circ \bar{\mathcal{Q}}(\mu, \bar{\zeta}, \partial_{\bar{\zeta}}) |p|^{2(\mu+1)} = b(\mu) \bar{b}(\mu) |p|^{2\mu}. \tag{5.62}$$

Formula (5.62) provides some quantitative information about the order and the growth of the distributions involved in the representation formula (5.60), when one identifies the Laurent development of both sides at $\mu = 0$. Note that for convexity reasons

$$2\operatorname{Re}(< \partial\tilde\varphi(\zeta),\, z - \zeta >) \leq \tilde\varphi(z) - \tilde\varphi(\zeta). \tag{5.63}$$

We profit from the inequality (5.63), and from the fact that in both kernels $\tilde P$ and $\tilde K$ appears the expression

$$\left(\frac{1+ < z,\bar\zeta >}{1 + \|\zeta\|^2}\right)^{N-n},$$

to choose N sufficiently large, to compensate the polynomial growth of every distribution involved in the analytic continuation. Note that the presence of the factor

$$\exp(2\operatorname{Re} < \partial\tilde\varphi(\zeta),\, z - \zeta >) \tag{5.64}$$

compensates the exponential growth, in the imaginary directions, of $\hat\phi$ (recall the inequality (5.58).) When N is sufficiently large and we let $R \to \infty$, we obtain, for any $z \in \mathbb{C}^n$, a division formula with remainder term for $\hat\phi(z)$, namely,

$$\begin{aligned}\hat\phi(z) =& P_1(z)\Phi_1(z) + \ldots + P_m(z)\Phi_m(z)\\ &+ \frac{(-1)^{\frac{m(m-1)}{2}}}{(2\pi i)^{n-m}} < \bar\partial(\frac{1}{P})(\zeta),\, \hat\phi(\zeta)g(z,\zeta) \wedge T_N(z,\zeta) >, \end{aligned} \tag{5.65}$$

where

$$T_N(z,\zeta) = \sum_{\alpha_1+\alpha_2=n-m} \frac{1}{\alpha_2!}\binom{N}{\alpha_1}e^{2<\partial\tilde\varphi,z-\zeta>}\left(\frac{1+ < z,\bar\zeta >}{1 + \|\zeta\|^2}\right)^{N-\alpha_1} \wedge (2\partial\bar\partial\tilde\varphi)^{\alpha_2}.$$

The form of the kernels $\tilde P$ and $\tilde K$, and the fact the exponential growth in z is controlled by (5.64), imply that the entire functions Φ_j are Fourier transforms of distributions Θ_j, supported by $\bar\Omega$. We now replace $\hat\phi$ by its representation (5.65) in the Plancherel formula (5.57). We observe that

$$\int_{\mathbb{R}^n} \widehat{f\chi}(\xi)P_j(\xi)\hat\Theta_j(\xi)d\xi = \int_{\mathbb{R}^n} \widehat{f\chi}(\xi)\mathcal{F}(P_j(-i\frac{\partial}{\partial x})\Theta_j)(\xi)d\xi$$

$$= (2\pi)^n < P_j(-i\frac{\partial}{\partial x})\Theta_j,\, f\chi >$$

$$= (2\pi)^n < \Theta_j,\, P_j(i\frac{\partial}{\partial x})(f\chi) >$$

$$= (2\pi)^n < \Theta_j,\, P_j(i\frac{\partial}{\partial x})f >= 0,$$

since $\mathrm{supp}(\Theta_j) \subseteq \bar{\Omega}$ and $\chi \equiv 1$ near $\bar{\Omega}$. Therefore, for any $\phi \in \mathcal{D}(\Omega)$ we have,

$$\int_{\mathbb{R}^n} f(x)\phi(x)dx \tag{5.66}$$

$$= (2\pi)^{-n} \frac{(-1)^{\frac{m(m-1)}{2}}}{(2\pi i)^{n-m}} \int_{\mathbb{R}^n} \widehat{f\chi}(\xi) < \bar{\partial}\frac{1}{P}(\varsigma), \, \hat{\phi}(\varsigma)g(\xi,\varsigma) \wedge T_N(\xi,\varsigma) > d\xi.$$

One can verify that, if N is sufficiently large, we are allowed to apply Fubini's theorem and the Lebesgue differentiation theorem (with respect to a parameter), to the integral (5.66). This is quite technical and it is based on explicit formulas for the residue current in terms of the Bernstein-Sato functional equation (5.51) for $p = P_1 \cdots P_m$. We will omit this point and refer to [75],[13]. We deduce that

$$\int_{\mathbb{R}^n} f(x)u(x)dx = \kappa(n,m) < (\hat{\phi}\bar{\partial}\frac{1}{P})(\varsigma), \int_{\mathbb{R}^n} \widehat{f\chi}(\xi)g(\xi,\varsigma) \wedge T_N(\xi,\varsigma)d\xi >$$

$$= \kappa(n,m) \int_{\mathbb{R}^n} \phi(t) < \bar{\partial}\frac{1}{P}(\varsigma), e^{-i<\varsigma,t>} \int_{\mathbb{R}^n} \widehat{f\chi}(\xi)g(\xi,\varsigma) \wedge T_N(\xi,\varsigma)d\xi > dt,$$

where we have denoted

$$\kappa(n,m) = (2\pi)^{-n} \frac{(-1)^{\frac{m(m-1)}{2}}}{(2\pi i)^{n-m}}.$$

So that we have in Ω the following representation formula

$$f(t) = \kappa(n,m) < \bar{\partial}\frac{1}{P}(\varsigma), e^{-i<\varsigma,t>} \int_{\mathbb{R}^n} \widehat{f\chi}(\xi)g(\xi,\varsigma) \wedge T_N(\xi,\varsigma)d\xi >. \tag{5.67}$$

The formula (5.67) is significantly more explicit than the representation formulas of Ehrenpreis and Palamodov, since their formulas are obtained by application of the Hahn-Banach theorem, so f is represented using a measure, about which one has very little knowledge. On the other hand, here we see the Fourier transform of f (or, at least, $f\chi$) appear. To counterbalance this observation, (5.67) is not really an integral representation, but the action of a current. More correctly, in the Ehrenpreis-Palamodov approach, one integrates against an abstract measure the exponential polynomials obtained by applying the Noetherian differential operators of the system (5.56) to the exponential function. Therefore, the question, which naturally arises, is what is the relation between the residue current associated to $P_1, \ldots, P_m$ and the Noetherian

operators. (For example, in [77] the Noetherian operators are computed using the Leray iterated residues of Chapter 3, §1.) This is really a question
about the structure of the residue current, which has been considered (in the
case of codimension one) by Coleff-Herrera, Dolbeault, Boudiaf [30], [35], [36].
Let us also mention that the Fundamental Principle of Eherenpreis-Palamodov
is valid for arbitrary systems (including matrices) of linear partial differential
equations with constant coefficients. Moreover, it applies not only to smooth
functions f but to distributions [37]. (For other proofs see [21].)

The preceding variation on the Fundamental Principle was considered in
[75], it allows for a representation valid on the whole set U, provided it is
convex, and even for distributions [13]. It has been extended to convolution
equations in $\mathbb{R}^n$ in [73]. The same method can be used, but with significantly
more technical difficulties, when dealing with slowly decreasing systems of convolution equations [16]. When the system (5.56) is hypoelliptic, besides being
normal, the formula (5.67) can be simplified by the fact that only the weight
q_3 is necessary, and one can use it to recover solutions of classical systems of
partial differential equations. For instance, for the system of Cauchy-Riemann
equations, one recovers the Henkin representation formulas in strictly convex
domains [19]. There are several applications where the explicit formula (5.67)
may be useful. One of them, seems to be the Fischer problem, namely, how
to construct a solution of the equation $P(\partial_z)f = 0$, with prescribed data on
another algebraic hypersurface $\{Q = 0\}$ in $\mathbb{C}^n$ [53]. This seems to be related
to the previously mentioned work of Leray on the Cauchy problem.

§4 The role of the Mellin transform

In this book we have proposed an approach to the notion of residue current that
was essentially based on the idea of analytic continuation of distributions of the
form $|f_1|^{\lambda_1} \cdots |f_p|^{\lambda_p}$. Since many problems remain open in this direction, we
would like to take the opportunity, in this brief final section, to mention some
of them as well as some (sometimes very) partial solutions. The first thing
we would like to emphazise here is the approach to the residue current using
Bochner-Martinelli type formulas. This approach has been one of our main
tools when applied to the cohomological residue in the sense of Dolbeault; a
careful reading of Chapters 4 and 5 shows that formulas of the form (3.32) and
(3.33) play an essential role in our reasonings. Since we have developed the
concept of residue current, it is natural to ask whether such formulas remain
true for arbitrary smooth forms, instead of $\bar{\partial}$-closed forms. It happens that
results obtained from the point of view of the analytic continuation of distributions, are easier to get than results involving integrals over cycles. We have
already remarked how the direct approach to the residue current as a limit of
integrals over cycles of the form Γ_ϵ could be difficult (compare, for example,

the need in [30] of admissible paths, as we mentioned in Remark 3.19, or the fact that, in his work, [57], [58], Passare needs to take averages of limits.)

Let us show first that our formulas (3.32) and (3.33) remain valid for any test function; we prove it here only in the simplest situation, that is, when the multiindex β is zero. The general case can be dealt with similarly [7],[14].

Proposition 5.21. *Let $f_1, \ldots, f_p$ be p holomorphic functions defining a complete intersection variety in some open subset $U \subset \mathbb{C}^n$. Let ϕ be a $(n, n-p)$ test form with coefficients in $\mathcal{D}(U)$. The function of one complex variable μ defined for $\operatorname{Re} \mu \gg 0$ as*

$$\mu \mapsto \mathcal{J}(\mu; \phi) = \mu \int |f_1 \ldots f_p|^{2\mu} \frac{\overline{\partial f_1} \wedge \ldots \wedge \overline{\partial f_p}}{\|f\|^{2p}} \wedge \phi$$

can be analytically continued as a meromorphic function in $\mathbb{C}$. The analytic continuation is holomorphic at the origin and the value at the origin equals

$$(-1)^{\frac{p(p-1)}{2}} (2\pi i)^p \frac{1}{p!} < \bar{\partial} \frac{1}{f}, \phi > . \tag{5.68}$$

Remark 5.22. In fact, the same statement has a multivariable analogous (and the proof is similar); the function of p complex parameters λ defined in $\operatorname{Re} \lambda_j \gg 0$, $j = 1, \ldots, p$, by

$$\lambda \mapsto \frac{1}{p}(\lambda_1 + \cdots + \lambda_p) \int |f_1 \ldots f_p|^{*2\lambda} \, \bar{f}^{*\beta} \frac{\overline{\partial f} \wedge \phi}{\|f\|^{2(p+|\beta|)}}$$

can be analytically continued in some orthant $\operatorname{Re} \lambda_j \geq -\gamma$, $j = 1, \ldots, p$, and its value at the origin equals

$$(2\pi i)^p (-1)^{\frac{p(p-1)}{2}} \frac{\beta!}{p(p-1+|\beta|)!} < \bar{\partial}(\frac{1}{f^{*(\beta+\underline{1})}}), \phi > . \tag{5.69}$$

(For a detailed proof, see [14], Theorem 2).

Proof of Proposition 5.21. The existence of the analytic continuation of $\mathcal{J}(\mu; \phi)$ can easily be proved if one combines a resolution of singularities π (relative to the product $f_1 \cdots f_p$) and the pull-back via the projections associated to toroidal manifolds corresponding to the monomials appearing in the expressions for $\pi^* f_1, \ldots, \pi^* f_p$ in the different charts of $\mathcal{X}$ (such a method has already been used in the proof of Theorem 3.25.) The fact that the analytic continuation is holomorphic near the origin follows of the same argument. When $p = n$, one can assume that the only common zero of the f_j in U is the origin; it is immediate to check, as in the proof of Theorem 3.25 and of Proposition 4.8, that the analytic continuation of $\mathcal{J}$, apart from the fact that it is holomorphic at the origin, is such that, for any j in $\{1, \ldots, n\}$, $\mathcal{J}(0, \bar{\zeta}_i \phi) = 0$. We already know (from the proof of Proposition 4.8), that the same results holds for the function

$$\mu \mapsto J(\mu; \phi) = \mu^n \int \frac{|f|^{*\mu 2}}{|f|^{*2}} \,\overline{\partial f} \wedge \phi.$$

Moreover, the value of $\mathcal{J}(0; \phi)$ has already been computed in (3.33) if ϕ is $\bar{\partial}$ closed near $\{f_1 = \ldots = f_n = 0\}$. These two remarks are sufficient to ensure the validity of the proposition when $p = n$. In fact, the two currents are supported at the origin and their action ignores the non-holomorphic terms of the Taylor expansion of a test function.

Our proof of the proposition in the general case will use an induction on the number $n - p$. We have just seen that the proposition is true when $n - p = 0$. Let us assume the inductive hypothesis and consider p functions defining a complete intersection in a neighborhood of the origin U; the test form ϕ will be fixed, with support in an arbitrary small neighborhood of the origin. The idea of our inductive proof is related to the method of fibered residues (that is, a method based on slicing), which has been extensively developed in [30], [72],[65].

If we rotate coordinates, we can assume that the collection $\zeta_1, f_1, \ldots, f_p$ defines an analytic set of dimension $n - p - 1$ in U and that for generic values ζ_1^0 of ζ_1, one has, for any ζ' close to the origin in $\mathbb{C}^{n-1}$,

$$\dim_{\zeta'}\{f_1(\zeta_1^0, \zeta') = \ldots = f_p(\zeta_1^0, \zeta') = 0\} \le n - p - 1.$$

We introduce functions K_1 and K_2, of two complex variables, defined first only for Re $\lambda_1 \gg 0$, Re $\lambda_2 \gg 0$, by

$$K_1(\lambda) = \lambda_1 \int |\zeta_1|^{2\lambda_2} |f_1 \ldots f_p|^{2\lambda_1} \frac{\overline{\partial f}}{\|f\|^{2p}} \wedge \phi$$

$$K_2(\lambda) = \frac{\lambda_1^p}{p!} \int |\zeta_1|^{2\lambda_2} |f_1 \ldots f_p|^{2(\lambda_1 - 1)} \overline{\partial f} \wedge \phi$$

As functions of two complex variables, K_1 and K_2 can be analytically continued to meromorphic functions in $\mathbb{C}^2$. In order to analyze the analytic continuation, we use a resolution of singularities for the hypersurface $f_1 \ldots f_p \zeta_1 = 0$, together with monoidal changes of coordinates corresponding to toroidal manifolds. In both cases, the polar set is a union of hyperplanes $\alpha\lambda_1 + \beta\lambda_2 + \gamma = 0$, where $\alpha, \beta \in \mathbb{N}$ and $\gamma \in \mathbb{Z}$. The test form ϕ can be written as

$$\phi = \phi_1 + d\bar{\zeta}_1 \wedge d\zeta_1 \wedge \psi = \phi_1 + \phi_2$$

where $d\bar{\zeta}_1$ does not appear in ϕ_1. The analytic continuation of K_2 as a function of two variables has already been considered in the proof of Theorem 3.18;

since ζ_1 and the f_j define a complete intersection, one can write the analytic continuation of K_2 in the form (3.40), that is,

$$K_2(\lambda) = h_2(\lambda) + \sum_\tau \frac{\lambda_1^p k_\tau(\lambda)}{\prod_k (\rho_{\tau,k}\lambda_1 + \sigma_{\tau,k}\lambda_2)}$$

where the products $\prod_k$ are products of at most $p-1$ factors of the form $\rho_{\tau,k}\lambda_1 + \sigma_{\tau,k}\lambda_2$; the constants $\rho_{\tau,k}$ and $\sigma_{\tau,k}$ are respectively in $\mathbb{N}^*$ and $\mathbb{N}$; the functions h_2 and k_τ are holomorphic near the origin in $\mathbb{C}^2$.

We claim that the value $h_2(0,0)$ does not change if one replaces in the expression of K_2, the differential form ϕ by ϕ_2; this follows from the fact that, in a local chart, $K_2(\lambda)$ can be expressed as a linear combination of terms of the form

$$\lambda_1^p \int |\pi^*\zeta_1|^{2\lambda_2}|w|^{*2\lambda_1 k} \frac{1}{w^{*l}} \frac{d\bar{w}_J}{\bar{w}_J} \wedge \theta(\lambda_1, w)\pi^*(\phi)$$

for some multiindex k, l, some subset J of $\{1, \ldots, n\}$ of cardinal less than p and some function θ which is smooth, with compact support in w, and depends holomorphically on λ_1. Now one can see that terms arising from the contribution of ϕ_1 are such that, at least one w_j, $j \in J$, is already in the monomial factorization of $\pi^*(\zeta_1)$. As seen in the proof of Theorem 3.18, the contribution to such terms to $h_2(0,0)$ is zero. On the other hand, it is easy to see that the function K_1 can be continued as an analytic function up to an open orthant containing the origin and that its value at 0 can be computed when replacing ϕ by ϕ_2, so that, in order to prove our result, that is $h_2(0,0) = K_1(0,0)$, one can replace ϕ by ϕ_2, which we will do from now on in the expressions of K_1 and K_2.

We consider $K_1(\lambda_1, \lambda_2^{(0)})$ and $K_2(\lambda_1, \lambda_2^{(0)})$ for $\operatorname{Re} \lambda_2^{(0)} \gg 0$. Using Fubini's theorem, for $\operatorname{Re} \lambda_2^{(0)} \gg 0$, $K_1(\lambda_1, \lambda_2^{(0)})$ can be written as

$$\pm \lambda_1 \int |\zeta_1|^{2\lambda_2^{(0)}} d\bar{\zeta}_1 \wedge d\zeta_1 \wedge \left(\int_{\zeta'} |(f_1 \ldots f_p)(\zeta_1, \zeta')|^{2(\lambda_1-1)} \frac{\overline{\partial_{\zeta'} f}}{\|f\|^{2p}} \wedge \psi \right).$$

The symbol $\int_{\zeta'}$ indicates the integration takes place on the variables ζ'. Similarly for $K_2(\lambda_1, \lambda_2^{(0)})$, one obtains

$$\pm \frac{\lambda_1^p}{p!} \int |\zeta_1|^{2\lambda_2^{(0)}} d\bar{\zeta}_1 \wedge d\zeta_1 \wedge \left(\int_{\zeta'} |(f_1 \ldots f_p)(\zeta_1, \zeta')|^{2(\lambda_1-1)} \overline{\partial_{\zeta'} f} \wedge \psi \right).$$

Fix a value $\zeta_1^0 \neq 0$ and let us study the analytic continuation of

$$\lambda_1 \mapsto \lambda_1 \int_{\{\zeta_1 = \zeta_1^0\}} |(f_1 \ldots f_p)(\zeta)|^{2\lambda_1} \frac{\overline{\partial_{\zeta'} f}}{\|f\|^{2p}} \wedge \psi(\zeta_1^0, \zeta') \qquad (5.70)$$

To deal with such a function of λ_1, we use Hironaka's theorem (relative to $\zeta_1 f_1 \ldots f_p$), and the toroidal embeddings $\tilde{\pi}$. In a local chart, where we have chosen a centered system of coordinates t, we have $(\tilde{\pi} \circ \pi)^* \zeta_1(t) = v(t) t^{*\gamma}$ for some invertible function v and some multiindex γ. (v can be incorporated into one of the coordinates t, and we shall do so from now on.) The analytic continuation of (5.70) can be viewed as the analytic continuation of the expression of the form

$$\lambda_1 \int_{\{t^{*\gamma}=\zeta_1^0\}} |\tilde{v}|^{2\lambda_1} |t|^{*2\lambda_1 \omega} \frac{\bar{m}^p \theta_1 + \theta_2 \bar{m}^{p-1} d\bar{m}}{|m|^{2p}} \wedge \sigma \qquad (5.71)$$

where m is a monomial (namely, the distinguished monomial, see the proof of Theorem 3.25), σ is a $(n-1, n-2)$ form, θ_1 and θ_2 two elements respectively in $\Lambda_c^{(0,1)}(\mathbb{C}^2)$ and $\mathcal{D}(\mathbb{C}^2)$, $\tilde{v}$ a non vanishing holomorphic function on $\mathrm{supp}(\theta_1) \cup \mathrm{supp}(\theta_2)$, and ω a multiindex. Polar coordinates can be used to obtain (as in [4]) the analytic continuation of (5.71), ζ_1^0 playing the role of a parameter. For fixed $\epsilon > 0$, only a finite number of poles appears in the half-space $\mathrm{Re}\ \lambda_1 \geq -\epsilon$. The polar parts, in the Laurent expansion at these poles, of the meromorphic continuation of (5.71) have coefficients bounded by $C/|\zeta_1^0|^M$ for some absolute constants M and C, independent of ζ_1^0; moreover, the set of poles is included in a subset of $\mathbb{Q}^*$, which is independent of ζ_1^0. The same conclusion remains valid for the analytic continuation of

$$\lambda_1 \mapsto \frac{\lambda_1^p}{p!} \int_{\{\zeta_1 = \zeta_1^0\}} |(f_1 \ldots f_p)(\zeta)|^{2(\lambda_1-1)} \overline{\partial_{\zeta'} f} \wedge \psi(\zeta_1^0, \zeta').$$

Therefore, using Lebesgue's dominated convergence theorem, one can see that for $\mathrm{Re}\ \lambda_2^{(0)} \gg 0$,

$$K_1(0, \lambda_2^{(0)}) = \pm \int |\zeta_1|^{2\lambda_2^{(0)}} d\bar{\zeta}_1 \wedge d\zeta_1 \wedge$$

$$\wedge \left(\lim_{\lambda_1 \to 0} \lambda_1 \int_{\zeta'} |(f_1 \ldots f_p)(\zeta_1, \zeta')|^{2\lambda_1} \frac{\overline{\partial_{\zeta'} f}}{\|f(\zeta_1, \zeta')\|^{2p}} \wedge \psi(\zeta_1, \zeta') \right), \qquad (5.72)$$

where the limit is understood as the value at the origin of the analytic continuation. The same reasoning shows that the value $K_1(0, \lambda_2^{(0)})$ equals

$$\pm \int |\zeta_1|^{2\lambda_2^{(0)}} d\bar{\zeta}_1 \wedge d\zeta_1 \wedge \left(\lim_{\lambda_1 \to 0} \frac{\lambda_1^p}{p!} \int_{\zeta'} |(f_1 \cdots f_p)(\zeta_1, \zeta')|^{2(\lambda_1-1)} \overline{\partial_{\zeta'} f} \wedge \psi(\zeta_1, \zeta') \right) \qquad (5.73)$$

Now, for any generic ζ_1^0, the p functions of $n-1$ variables

$$\zeta' \mapsto f_j(\zeta_1^0, \zeta')$$

define an analytic variety of dimension $n-1-p$ in a neighborhood of the origin in $\mathbb{C}^{n-1}$, and we can apply the inductive hypothesis to conclude that the two limits in the integrands (5.72) and (5.73) are equal for almost all ζ_1. We conclude that, for Re $\lambda_2^{(0)} \gg 0$, the values of $K_1(0, \lambda_2^{(0)})$ and $K_2(0, \lambda_2^{(0)})$ coincide; the proof of the equality $h_2(0,0) = K_1(0,0)$ follows using analytic continuation. $\qquad\square$

This method can also be used to prove the Transformation Law for systems defining a complete intersection (see Theorem 4.9 for the discrete case.) In a recent preprint [33] there is a different proof, based on the method of Herrera of fibered residues.

Theorem 5.23. *Let $f_1, \ldots, f_p, \varphi_1, \ldots, \varphi_p$ be two sets of p holomorphic functions in an open set $U \subseteq \mathbb{C}^n$, each of them defining a complete intersection in U. Suppose that for $1 \le j \le p$ one has*

$$\varphi_j = \sum_{k=1}^p a_{j,k} f_k$$

for some holomorphic functions $a_{j,k}$ and let $\Delta = \det[a_{j,k}]$. Then, for any test form ϕ of type $(n, n-p)$ in U,

$$< \bar\partial \frac{1}{f}, \phi > = < \bar\partial \frac{1}{\varphi}, \Delta\phi > .$$

Proof. One repeats the inductive argument from the proof of Proposition 5.21, since we already know the result is true for $n-p=0$. In a neighborhood of the origin, choose a convenient system of coordinates and replace K_1 and K_2 in the preceding proof, by the two functions of two complex variables

$$K_1(\lambda) := \lambda_1^p \int |\zeta_1|^{2\lambda_2} |f_1 \cdots f_p|^{2(\lambda_1-1)} \overline{\partial f} \wedge \phi.$$

$$K_2(\lambda) := \lambda_1^p \int |\zeta_1|^{2\lambda_2} |\varphi_1 \cdots \varphi_p|^{2(\lambda_1-1)} \Delta \overline{\partial \varphi} \wedge \phi.$$

The rest of the proof is the same. $\qquad\square$

In fact, statements like Proposition 5.21 correspond to different Tauberian approaches to recover the value of the action of the residue current. This was our idea when proving formulas (3.32) and (3.33), in the case the test forms were $\bar\partial$-closed near the set of common zeroes. These procedures are not unique;

to give another example, one can check in the discrete case, that the action of the residue current on a test form can be obtained as

$$\lim_{\epsilon \to 0} \frac{(-1)^{\frac{n(n-1)}{2}} n!}{(2\pi i)^n} \epsilon \int \frac{\overline{\partial f}}{(\|f\|^2 + \epsilon)^{n+1}} \wedge \phi.$$

Any of these differents results seems to indicate that, when $f_1, \ldots, f_p$ define a complete intersection in U and ϕ is a $(n, n-p)$ test form with support in U, the almost everywhere defined function

$$\epsilon \mapsto I(\phi; \epsilon) = \int_{\Gamma_f(\epsilon)} \frac{\phi}{f_1 \cdots f_p} \tag{5.74}$$

has a limit when ϵ tends to $\underline{0}$ in $(\mathbb{R}^+)^p$. This seems to be a very interesting question, which could be studied via our methods of analytic continuation. In fact, even when $p = 1$, we have already noticed in Remark 3.5 that the existence of the limit (5.74) was not trivial, and could not be deduced from the Weierstrass Preparation Theorem.

Let us explain an idea, which is due to Barlet and Björk, that shows the existence of the limit without using directly resolution of resolution of singularities. As we have said several times, the Mellin transform of $\epsilon \mapsto I_f(\phi, \epsilon)$ is the function

$$J : \mu \mapsto \mu \int |f|^{2(\mu-1)} \overline{\partial f} \wedge \phi$$

The fact that this function is holomorphic up to the origin, can be seen to be a consequence of Kashiwara's theorem [47], showing that the roots of the Bernstein-Sato polynomial in (5.51) are strictly negative rational numbers. Let us write the functional equation, after one iteration, as follows

$$B(\mu)B(\mu+1)|f|^{2\mu}\frac{1}{f} = \bar{Q}(\mu, \bar{\zeta}, \partial_{\bar{\zeta}})\left(\frac{|f|^{2\mu} f^2}{f}\right)$$

and remark that, using Stokes's theorem, we can write J as

$$J(\mu) = -\int \frac{|f|^{2\mu}}{f} \bar{\partial}\phi$$

This expression can be further rewritten using integration by parts, via the functional equation, as

$$-\frac{1}{B(\mu)B(\mu+1)} \int |f(\zeta)|^{2\mu} \left(\frac{\bar{f}}{f}\right) \bar{f}\psi, \tag{5.75}$$

for some smooth form ψ. The holomorphy at the origin follows from the previously recalled property of the Bernstein polynomial. On the other hand,

Kashiwara proved in [47] (and that can be also be obtained, as in [5], as a consequence of the algebraic dependence of f on its Jacobian ideal, which we already quoted in Chapter 3, §5), that there exists a functional equation of the form

$$\mu^N f^\mu = \sum_{j=1}^{N} \mu^{N-j} A_j(\zeta, \partial)(f^\mu) \tag{5.76}$$

where the differential operator A_j has order at most j. The identity (5.76) can be used to show that the function J is rapidly decreasing on vertical lines. The continuity of I, its antecedent via Mellin transform, follows from classical results about the inversion of the Mellin transform [34].

In the case $p = 2$, we have shown during the proof of Theorem 3.18 that the Mellin transform of I is holomorphic at the origin as a function of two variables. In particular cases, when we have an analogue of the functional equation (5.76) (e.g., when the f_j are homogeneous and thus, we can use the Euler identity, and find a relation of the form (5.76) for the product $f_1^{\lambda_1} f_2^{\lambda_2}$ and let $N = 1$) it is possible to show the rapid decrease of J on planes Re $\lambda_1 = c_1$, Re $\lambda_2 = c_2$, and therefore obtain the continuity of I. (There is some very recent work of Passare and Tsikh on this type of questions [59].)

To conclude, we hope to have convinced the reader of the importance of the multidimensional Grothendieck residue as a powerful tool in Analysis, Commutative Algebra, Number Theory, and even in Computer Algebra. Furthermore, we wish we made clear that a number of deep questions about this theory remain unsolved, and attracted the interest of the reader on them.

References for Chapter 5

[1] F. Amoroso, Tests d'appartenance d'après un théorème de Kollár, Comptes Rendus Acad. Sci. Paris Ser. I 309(1989), 691–694.

[2] F. Amoroso, Membership problem, in *Diophantine Approximations and Transcendental Numbers, Luminy 1990*, P. Philippon (ed.), Walter de Gruyter, Berlin 1992, 15–37.

[3] F.Amoroso, On a conjecture of C. Berenstein and A. Yger, preprint.

[4] D. Barlet, Développements asymptotiques des fonctions obtenues par intégration sur les fibres, Invent. Math. 68(1982), 129–174.

[5] D. Barlet and H.M. Maire, Développements asymptotiques, Transformation de Mellin Complexe et intégration sur les fibres, in Lect. Notes in Math. 1295, Springer-Verlag, New York, 1987.

[6] D.Bayer and M. Stillman, On the complexity of computing syzygies, J. Symbolic Comp. 6(1988), 135–147.

[7] C.A. Berenstein, R. Gay, and A. Yger, Analytic continuation of currents and division problems, Forum Math.1(1989), 15–51.

[8] C.A.Berenstein and D.C. Struppa, Complex Analysis and Convolution equations, *Contemporary Problems in Mathematics. Fundamental Directions*, Vol 54, VINITI, Moscow, 1990 (in Russian), English version to be published by Springer-Verlag.

[9] C.A. Berenstein and B.A. Taylor, Interpolation problems in $\mathbb{C}^n$ with applications to harmonic analysis, J. Analyse Math. 38(1980) 188–254.

[10] C.A. Berenstein and A. Yger, Analytic Bezout Identities, Advances in Applied Math. 10(1989) 51–74.

[11] C.A. Berenstein and A. Yger, Bounds for the degrees in the division problem, Mich. Math. J. 37(1990) 26–44.

[12] C.A. Berenstein and A. Yger, Effective Bezout identities in $\mathbb{Q}[z_1, \ldots, z_n]$, Acta Mathematica, 166(1991), 69–120.

[13] C.A. Berenstein and A. Yger, About Ehrenpreis' Fundamental Principle, in *Geometric and Algebraic Aspects in Several Complex Variables*, C.A. Berenstein and D.C. Struppa (ed.), Editel, Cosenza, 1991.

[14] C.A. Berenstein and A. Yger, Une formule de Jacobi et ses conséquences, Ann. Sci. Ec. Norm. Sup. Paris 24(1991), 363–377.

[15] C.A. Berenstein and A. Yger, Formules de représentation intégrale et problèmes de division, in *Diophantine Approximations and Transcendental Numbers, Luminy 1990*, P. Philippon (ed.), Walter de Gruyter, Berlin 1992, 15–37.

[16] C.A. Berenstein and A. Yger, Exponential polynomials and D-modules, preprint, 1992.

[17] B. Berndtsson and M. Andersson, Henkin-Ramirez formulas with weight factors, Ann. Inst. Fourier 32(1982), 91–110.

[18] B. Berndtsson, A formula for interpolation and division in $\mathbb{C}^n$, Math. Ann. 263(1983), 399–418.

[19] B. Berndtsson and M. Passare, Integral formulas and explicit version of the Fundamental Principle, J. Fcnal. Anal. 84(1989), 358–372.

[20] I.N. Bernstein, The analytic continuation of generalized functions with respect to a parameter, Funct. Anal. Appl. 6(1972), 273–285.

[21] J.-E. Björk, *Rings of Differential Operators*, North-Holland, Amsterdam 1979.

[22] A. Borel and al., *Algebraic D-Modules*, Academic Press, Inc., Orlando, Florida, 1987.

[23] W.D. Brownawell, Bounds for the degrees in the Nullstellensatz, Annals of Math. 126(1987) 577–591.

[24] W.D. Brownawell, Local diophantine Nullstellen inequalities, J. Amer. Math. Soc. 1(1988), 311–322.

[25] W.D. Brownawell, A prime power version of the Nullstellensatz, Michigan Math. J., to appear.

[26] B. Buchberger, An algorithmic method in polynomial ideal thory, in *Multidimensional Systems Theory*, (ed N.Bose), Reidel Publ, Dordrecht, 1985.

[27] L. Caniglia, A. Galligo, and J. Heintz, Borne simple exponentielle pour les degrés dans les théorèmes de zéros sur un corps de caractéristique quelconque, Comptes Rendus Acad. Sci. Paris, Ser. I 307(1988), 255–258.

[28] L. Caniglia, A. Galligo, and J. Heintz, Some new effective bounds in computational geometry, in Proc. AAECC-6, Lecture Notes in Computer Science 357, Springer-Verlag, Berlin, 1988.

[29] L. Caniglia, A. Galligo, J. Heintz, Equations for the projective closure and effective Nullstellensatz, Discrete Applied Mathematics 33(1991), 11–23.

[30] N. Coleff and M. Herrera, *Les courants résidus associés à une forme méromorphe*, Lect. Notes in Math. 633, Springer-Verlag, New York, 1978.

[31] A. Dickenstein and C. Sessa, An effective residual criterion for the membership problem in $\mathbb{C}[z_1, \ldots, z_n]$, J. of Pure and Applied Algebra 74(1991), 149–158.

[32] A. Dickenstein, N. Fitchas, M. Giusti, C. Sessa, The membership problem for unmixed polynomial ideals is solvable in single exponential time, Discrete Applied Mathematics 33(1991), 73–94.

[33] A. Dickenstein, La loi de transformation pour les courants résiduels associés à des intersections complètes manuscript, 1992.

[34] G. Doetsch, *Introduction to the Theory and Applications of the Laplace Transform*, Springer-Verlag, New York, 1974.

[35] P. Dolbeault, Sur la structure des courants résiduels, Revue Roumaine de Mathématiques XXXIII, 31–37.

[36] P. Dolbeault, On the structure of residual currents, preprint.

[37] L. Ehrenpreis, *Fourier Analysis in several Complex Variables*, Wiley-Interscience, New York, 1970.

[38] M. Elkadi, Une version effective du théorème de Briançon-Skoda dans le cas algébrique discret, preprint, Bordeaux, 1992.

[39] M. Elkadi, Bornes pour les degrés et les hauteurs dans le problème de division, preprint, Bordeaux, 1992.

[40] E. Fortuna and S. Lojasiewicz, Sur l'algébricité des ensembles analytiques complexes, J. reine angew. Math. 329(1981), 215–220.

[41] M. Giusti, Some effectivity problems in polynomial ideal theory, Eurosam 84, Lecture Notes in Computer Science 174, 159–171.

[42] D.I. Gurevich, Counterexamples to a problem of L. Schwartz, Funct. Anal. Appl. 9(1975), 116–120.

[43] G. Hermann, Die Frage derendliche vielen Schritte in der Theorie der Poly-nomideale, Math. Ann. 95(1926), 737–788.

[44] W.V.D. Hodge, D. Pedoe, *An Introduction to Algebraic Geometry, 3 Vols*, Cambridge Univ. Press, Cambridge, 1952.

[45] L. Hörmander, *The Analysis of Linear Partial Differential Operators, vol. II*, Springer-Verlag, New York, 1983.

[46] S. Ji, J. Kollár, B. Shiffman, A global Lojasiewicz inequality for algebraic varieties, Trans. Amer. Math. Soc., to appear.

[47] M. Kashiwara, B-functions and holonomic systems. Rationality of roots of b-functions, Invent. Math. 38(1976), 33–54.

[48] J. Kollár, Sharp effective Nullstellensatz, Journal of Amer. Math. Soc. 1(1988), 963–975.

[49] B. Lichtin, Generalized Dirichlet series and B-functions, Compositio Math. 65(1988), 81–120.

[50] F.S. Macaulay, *The algebraic theory of modular systems*, Cambridge Univ. Press, Cambridge, 1916.

[51] D.W. Masser and G. Wüstholz, Fields of large transcendence degree gener-ated by values of elliptic functions, Inventiones. Math. 72(1983), 407–464.

[52] E. Mayr and A. Meyer, The complexity of the word problem for com-mutative semi groups and polynomial ideals, Adv. in Math. 127(1988), 305–329.

[53] A. Méril and A. Yger, Problèmes de Cauchy globaux, Bull. Soc. Math. France (1991)

[54] Yu.V. Nesterenko, Bounds for the characteristic function of a prime ideal, Mat. Sbornik 123(1984), 11–34 = Math. USSR Sbornik 51(1985), 11–32.

[55] Yu.V.Nesterenko, On the measure of algebraic independence of the values of several functions, Math. USSR Sbornik 56(1986), 545–567.

[56] V.P. Palamodov, *Linear Differential Operators with Constant Coefficients*, Springer-Verlag, New York, 1970.

[57] M. Passare, Residues, currents, and their relation to ideals of holomorphic functions, Math. Scand. 62(1988), 75–152.

[58] M. Passare, A calculus for meromorphic currents, J. reine angew. Math. 392(1988), 37–56.

[59] M. Passare and A.K. Tsikh, Residue integrals and their Mellin Transforms, preprint.

[60] P. Philippon, Critères pour l'indépendance algébrique, Publ. Math. IHES 64(1986), 5–52.

[61] P. Philippon, A propos du texte de W.D. Brownawell: "Bounds for the degrees in the Nullstellensatz", Ann. of Math. 127(1988), 367–371.

[62] P. Philippon, Théorème des zéros effectif, d'après Kollár, Publ. Math. de l'Univ. P.et M.Curie, Paris VI, 88, Groupe d'étude sur les Problèmes diophantiens 1987–88, exposé 6.

[63] P. Philippon, Dénominateurs dans le théorème des zéros de Hilbert, Acta Arithmetica. 58(1991), 1–25.

[64] P. Philippon, Sur des hauteurs alternatives I, Math. Ann. 289(1991), 255–283.

[65] J.B. Poly, Formule de résidus et intersection des chaines sous-analytiques, Thesis, Université de Poitiers, 1974.

[66] C. Sabbah, Proximité évanescente I, Compositio Math. 62(1987), 283–328.

[67] C. Sabbah, Proximité évanescente II. Equations fonctionnelles pour plusieurs fonctions analytiques, Compositio Math. 64(1987), 213–241.

[68] B. Shiffman, Degree bounds for division problem in polynomial ideals, Michigan Math. J. 36(1989), 163–171.

[69] N. Sidiropoulos, M.S. EE. thesis, University of Maryland, College Park, MD, 1991.

[70] H. Skoda, Applications des techniques L^2 à la théorie des idéaux d'une algèbre de fonctions holomorphes avec poids, Ann. Sci. Ec. Norm. Sup. Paris 5 (1972), 545–579.

[71] B. Teissier, Résultats récents d'Algèbre commutative effective, Séminaire Bourbaki, 1989–1990, exposé 718, Astérisque 1990.

[72] A.K. Tsikh, *Multidimensional Residues and Their Applications*, Transl. Amer. Math. Soc. 103, 1992.

[73] A. Vidras, Ph.D. thesis, University of Maryland, College Park, MD, 1992.

[74] B.L. van der Warden, *Modern Algebra*, Springer-Verlag, New York, 1979.

[75] A. Yger, Formules de division et prolongement méromorphe, in Lect. Notes in Math. 1295, Springer-Verlag, New York, 1987.

[76] O. Zariski and P. Samuel, *Commutative Algebra*, Springer-Verlag, New York, 1958.

[77] D. Zeilberger, A new proof of the semilocal quotient structure theorem, Amer. J. Math. 100(1978), 1317–1332.

Progress in Mathematics

Edited by:

J. Oesterlé
Départment de Mathématiques
Université de Paris VI
4, Place Jussieu
75230 Paris Cedex 05, France

A. Weinstein
Department of Mathematics
University of California
Berkeley, CA 94720
U.S.A.

Progress in Mathematics is a series of books intended for professional mathematicians and scientists, encompassing all areas of pure mathematics. This distinguished series, which began in 1979, includes authored monographs, and edited collections of papers on important research developments as well as expositions of particular subject areas.

We encourage preparation of manuscripts in such form of TeX for delivery in camera-ready copy which leads to rapid publication, or in electronic form for interfacing with laser printers or typesetters.

Proposals should be sent directly to the editors or to: Birkhäuser Boston, 675 Massachusetts Avenue, Cambridge, MA 02139, U.S.A.